NUTRITION AND CANCER

An International Journal

This journal is abstracted or indexed in *Biosciences Information Services; Cambridge Scientific Abstracts: Health & Safety Science Abstracts, Risk Abstracts; Cumulative Index to Nursing and Allied Health Literature; CINAHL Database; EMBASE/Excerpta Medica; Index Medicus/ MEDLINE; ISI: Science Citation Index, Research Alert, SciSearch, Current Contents/Life Sciences, Current Contents/Clinical Medicine.*

First published 2001 by Lawrence Erlbaum Associates, Inc.

Published 2018 by Routledge
2 Park Square, Milton Park, Abingdon, Oxon OX14 4RN
52 Vanderbilt Avenue, New York, NY 10017

Routledge is an imprint of the Taylor & Francis Group, an informa business

ISBN 13: 978-0-8058-9681-7 (pbk)
ISSN 0163-5581
CODEN:NUCADQ

W0234963

NUTRITION AND CANCER, *40*(1), 1–2

IN MEMORIAM

Larry C. Clark
10 October 1948–20 March 2000

On Monday, 20 March 2000, Larry C. Clark, M.P.H., Ph.D., 51, Associate Professor of Epidemiology, College of Medicine, University of Arizona, died after a five-year battle with prostate cancer. It is a sad irony indeed that Larry developed this disease while conducting a trial that may well hold the key to its prevention. This trial involved selenium, an element that had been shown to be essential in 1957 but, from the late 1960s onward, had gained increasing recognition as a cancer-protecting agent. By the mid-1970s, animal experiments and comparisons of the selenium intakes with the cancer mortalities in different countries already strongly suggested that supplemental selenium at a safe nutritional dosage should significantly reduce human cancer risk, but to prove this, well-designed, long-term, large-scale human supplementation trials were needed. Larry Clark decided to design and conduct such a trial, for which he was ideally qualified.

Larry Clark's professional career began at the University of Michigan (Ann Arbor, MI), from which he received his M.P.H. in epidemiology in 1974. He subsequently moved to the University of North Carolina (Chapel Hill, NC), where, while holding appointments at the Institute for Environmental Studies and the Biometry Branch of the National Institute of Environmental Health Sciences (Research Triangle Park, NC), he continued his graduate studies in the Department of Epidemiology. In 1982, he obtained his doctorate in epidemiology with a dissertation entitled *A Case-Control Study of*
Skin Neoplasms and the Anticarcinogenic Effects of Selenium, which presaged his own future involvement with selenium and cancer prevention.

In 1982, Larry was appointed Assistant Professor of Epidemiology jointly at the Department of Preventive Medicine of the School of Veterinary Medicine and the Division of Nutritional Sciences of Cornell University, where he conceived his now-famous nutritional cancer prevention study, for which he was to receive the Pioneer Science Award of the Cancer Treatment Research Foundation in 1997.

The original aim of Clark's multicentric, fully randomized, and placebo-controlled study was to determine whether a daily supplement of 200 µg of selenium in a nutritional form was effective in preventing basal cell and squamous cell skin cancers in high-risk subjects, i.e., patients with a history of basal cell or squamous cell carcinoma of the skin during the preceding year. In the course of the study, a total of 1,312 eligible patients (mean age 63 yr, range 18–80 yr) were randomized from 1983 through 1991. After a total follow-up of 8,271 person-yr, the study seemed to end in failure: Selenium at this dosage was found to have no significant effect on basal cell or squamous cell skin cancer in these subjects. However, when, in 1990, funding became available for long-term follow-up, the selenium supplementation was found to be associated with significant reductions of total cancer incidence (all sites combined), lung, colorectal, and prostate cancer incidences, and lung cancer mortality. Specifically, selenium reduced the incidence of cancers of the prostate by 63%, of the lung by 58%, and of the colon by 46%.

In the light of these findings, Larry Clark decided to halt the study two years early in order not to unnecessarily delay the publication of these exciting results. Larry had yet another reason for concluding his study sooner than intended: Suffering from prostate cancer since the mid-1990s, his health was failing, and by December 1996, at the same time the results of his study were published in the *Journal of the American Medical Association*, he received the news that his condition was terminal. Undaunted and brave, he pressed on with his research, initiating new clinical trials and publishing papers. He explored ways of halting the progression of his disease and, yes, also incorporated selenium into his treatment, and while it is impossible to say whether selenium specifically helped him, Dr. David Alberts, his colleague at the University of Arizona Cancer Center, attests that he had "a very prolonged remission considering how severe his cancer was at diagnosis." His dedication to the cause remained unchanged, even in the face of impending defeat. He continued to write papers and submit innovative grant proposals in the fields of cancer prevention, nutrition, and clinical trials. Even though he was fully aware that he had only a few more months to live, he made every effort to secure the necessary grant support so that ongoing research projects at the Arizona Cancer Center could continue. On 21 February 2000, he mobilized his last remaining bodily reserves to attend a site visit to the University of Arizona Cancer Center by the National Institutes of Health. Returning to his sick bed, he was not to leave it again until his death. Larry will be sadly missed, but his memory will live on and will forever be linked to one of the most important studies in nutritional cancer prevention ever to be conducted.

Gerhard N. Schrauzer
Department of Chemistry and Biochemistry
University of California, San Diego

NUTRITION AND CANCER, *40*(1), 3

INTRODUCTION

This issue is a new departure for the *Journal*, in that it is devoted to a single topic, namely, the cancer-chemopreventive effects of the trace element selenium. Although epidemiological and animal model studies have contributed enormously to the development of this field, the clinical trial headed by the late Dr. Larry Clark brought to light the very real possibility that selenium compounds may serve as protective agents in populations at risk for prostate, colon, and lung cancers. For this reason, we have brought together experts from various disciplines to address the current state of our knowledge of the role of selenium as an anticancer agent. It is hoped that, by bringing these various approaches together in one place, the research community, both graduate students and established investigators, can better grasp the complex nature of this field.

The papers in this issue cover the entire spectrum of cancer research, ranging from clinical trials to animal model studies and molecular biology. In an overview, El-Bayoumy traces the historical development of the relationship between selenium and cancer, showing that selenium can, under specific conditions, be toxic, an essential nutrient, and an anticarcinogenic and antiangiogenic agent. Combs discusses the paradigm shift from the concept of nutrient sufficiency to that of optimal nutrition as it applies to chronic disease prevention. Medina and collaborators review studies on the chemopreventive effects of inorganic and organic forms of selenium on rodent mammary tumorigenesis and summarize the rationale for use of a recently discovered organic selenium compound (Se-methylselenocysteine) in human populations. In a similar vein, El-Bayoumy and colleagues describe their pioneering studies on synthetic organic selenium compounds, such as 1,4-phenylenebis(methylene)selenocyanate, and their chemopreventive effects in laboratory animal models of lung, oral cavity, colon, and breast cancer. Collectively, these organoselenium compounds, on the basis of both in vitro and in vivo studies, have been shown to display chemopreventive effects with significantly less toxic side effects than the historically used inorganic forms, such as sodium selenite.

Mechanistic studies are designed to provide biologically plausible explanations as to how a trace element, such as selenium, can act as an anticancer agent. Youn et al. report on molecular studies suggesting that selenium compounds may exert protective effects by binding to nuclear regulatory transcription factors such as nuclear factor-κB. Spallholz et al. describe their in vitro studies indicating that metabolism of L-selenomethionine to methylselenol may result in oxidative changes that trigger cellular apoptosis. Focusing again on the role of selenium metabolites, Fleming et al. report on their studies in oral carcinomas on the role of the natural selenium metabolite selenodiglutathione on the intracellular signaling pathways responsible for cell proliferation and apoptosis. Using primary cell culture systems, they show that selenodigluathione induces key mediators of apoptosis, including Fas ligand and the Jun NH_2-terminal kinase. Kim and Milner review a variety of cellular targets of selenium compounds, including enzymes involved in carcinogen bioactivation, cell proliferation check points, and apoptosis, and Gopalakrishna and Gundimeda detail their intriguing studies indicating that selenium-containing compounds can inhibit protein kinase C, a critical intracellular signaling molecule, by reacting with the sulfur groups on the cysteine-rich region of the catalytic domain of protein kinase C.

Neoangiogenesis plays a critical role in tumor development, since without new capillary growth, tumors cannot continue to grow and disseminate. Hence, blocking the steps involved in tumor angiogenesis is a viable strategy in preventing tumor metastasis. Lu and Jiang report on studies indicating that methylselenol and its metabolites act to block capillary endothelial cell migration into mammary tumors via inhibitory effects at the level of matrix metalloproteinases and tumor vascular endothelial growth factor production.

To translate this knowledge to the human setting, Marshall describes several clinical trials currently under way designed specifically to evaluate whether selenium supplementation can reduce the incidence of prostate cancer in high-risk populations.

I had the pleasure of meeting Larry Clark at a symposium in Greece and remember his intelligence, good humor, and willingness to explain epidemiology to amateurs like me. It was indeed a cruel irony that prostate cancer took his life, and it is hoped that his legacy will live on in the work of others and, especially, the contributors to this issue of the *Journal*.

Leonard A. Cohen, Ph.D.
Editor

NUTRITION AND CANCER, 40(1), 4–5

Overview: The Late Larry C. Clark Showed the Bright Side of the Moon Element (Selenium) in a Clinical Cancer Prevention Trial

The trace element selenium was discovered by the Swedish chemist Jöns J. Berzelius in 1818. As a naturally occurring element, selenium ranks 17th; its geographic distribution varies from high concentrations in the soil in certain regions of China, the former USSR, Venezuela, and the United States to rather low levels in New Zealand and Finland.

Early reports identified selenium as a highly toxic element. In fact, the toxicity of selenium in animals was first observed by Marco Polo during his journey from Venice to China in the 13th century, when some feed-related toxicity caused degenerative hoof disease. At that time, the toxic component in the animal feed was not known. In 1857, T. C. Madison, an army surgeon, noted a similar toxic effect, accompanied by other symptoms in horses at Fort Randall, SD. It was not until 1934 that the toxic component in the plants fed to horses was identified as a selenium compound.

An early report by Nelson in 1944 indicated an elevation in the number of liver adenomas in rats fed a diet enriched in seleniferous corn and wheat or inorganic selenide. When Nelson's experiment was repeated under carefully controlled conditions, Clayton and Baumann in 1949 and Harr in 1972 not only failed to confirm the earlier findings but, on the contrary, provided evidence that selenium can act as a chemopreventive agent in the liver. Close examination and reevaluation of the early investigations indicated that the animals did not have true neoplastic lesions in the liver but, rather, extensive regenerating nodules, concurrent with severe liver cirrhosis stemming from selenium toxicity.

Selenium was considered a toxic element until 1951. At that time, Klaus Schwartz, a physician from Germany working at the US National Institutes of Health as a research fellow, demonstrated the protective effect of vitamin E on liver necrosis in rats. He was the first to distinguish between "fatty liver and cirrhosis" and acute (massive) liver necrosis. His pioneering work in 1957 indicated that a component containing selenium (he named it factor 3) was an essential nutrient when it was found to protect laboratory animals from liver necrosis. In addition, he found that factor 3 was more potent than vitamin E in protecting the liver from necrosis.

Douglas V. Frost was the first to challenge the mistaken evidence of carcinogenicity induced by selenium. In 1969, he and Shamberger reported an epidemiological study suggesting that selenium might prevent, rather than cause, cancer. In fact, this epidemiological observation by Frost and Shamberger stimulated great research interest in the role of selenium on cancer prevention. Gerhard N. Schrauzer and his team, in an independent study in 1971, reported that selenium was a potential human cancer-protective agent. Further support for the protective role of selenium came in 1973 through the discovery by Rotruck and colleagues of the biochemical basis for an essential role of selenium in the functions of glutathione peroxidase, an enzyme responsible for preventing damage due to oxidative stress. Additional support for the protective role of selenium was based on numerous epidemiological studies conducted in the United States and abroad, as well as in preclinical investigations.

In preclinical studies and at the beginning of the 1970s, most of the investigations that have examined the role of selenium as a cancer-chemopreventive agent have utilized inorganic forms of the element. Selenium supplementation in the diet or drinking water has been shown to inhibit the development of neoplasms in liver, skin, pancreas, colon, and mammary glands. In addition, it has been shown that selenium can inhibit the initiation and postinitiation phases of chemical carcinogenesis. In the absence of chemical carcinogen treatments, selenium inhibited the formation of "presumably" virally induced mammary tumors. At the beginning of the 1980s, investigations were aimed at the development of novel synthetic organoselenium compounds and at the discovery of naturally occurring selenium compounds that are more effective and less toxic than inorganic forms of selenium. The outcomes of these studies were highly promising, and this area of research is growing rapidly. Exciting recent studies show that selenium is an antimetastatic and antiangiogenic agent. This indicates that it can be used as a chemopreventive as well as a chemotherapeutic agent.

The use of selenium in human clinical trials is limited. Such interventions have been conducted during the past decade in China, India, Italy, and the United States, with the element in the form of selenium-enriched yeast, selenite, or selenate. In certain trials, it was difficult to define the form of selenium that was given. Populations having different risk factors were recruited for these trials. Some, but not all, of the studies performed in China suffered from methodological problems, such as lack of quality controls. One of the most exciting clinical trials in the United States, conducted by the late Larry Clark, supported a protective effect of selenium-enriched yeast against cancer of the prostate, colon, and lung. The outcome of Clark's trial stimulated the initiation of two new clinical intervention trials in three European countries (PRECISE) and in the United States (SELECT). The pioneering work of the late Larry Clark, who passed away on 20 March 2000 from complications related to prostate cancer, resulted in highly useful information that helps scientists in this field progress toward the prevention of the disease that took Larry Clark's life. Larry Clark showed the bright side of the moon element in an important human clinical intervention trial. He has contributed nationally and internationally to the field of selenium and cancer prevention research. As a scientist in this field and as a friend to Larry Clark, I felt obligated to approach several colleagues during the American Association for Cancer Research meeting in San Francisco, CA, on 1–5 April 2000 and ask them to contribute to this issue of *Nutrition and Cancer*, which is dedicated to the memory of Larry Clark. All were happy to participate and responded on short notice in tribute to Dr. Larry C. Clark, to whom we owe a debt of gratitude for his outstanding discovery.

I express my deep appreciation to all contributors and, above all, Dr. Leonard Cohen, Editor, *Nutrition and Cancer*, for his positive reply to my idea of having a volume dedicated to Dr. Clark on the role of selenium in cancer prevention that ranges from epidemiological, molecular, and clinical interventions. At the same time, I apologize to the many researchers who have also contributed to this field but, because of time constraints, could not be invited to participate.

References

Clark LC, Combs GF, Turnbull BW, Slate EH, Chalker DK, et al.: Effects of selenium supplementation for cancer prevention in patients with carcinoma of the skin: a randomized controlled trial. *JAMA* **276**, 1957–1963, 1996.

El-Bayoumy K: The role of selenium in cancer prevention. In *Cancer: Principles and Practice of Oncology*, 4th ed, DeVita VT Jr, Hellman S, and Rosenberg SA (eds). Philadelphia: Lippincott, 1991, vol 2, pp 1–5.

El-Bayoumy K: The protective role of selenium on genetic damage and on cancer. *Mutat Res* **475**, 123–139, 2001.

Karam El-Bayoumy, Ph.D.
Division of Cancer Etiology and Prevention
1 Dana Road
American Health Foundation
Valhalla, NY 10595

NUTRITION AND CANCER, *40*(1), 6–11

Impact of Selenium and Cancer-Prevention Findings on the Nutrition-Health Paradigm

Gerald F. Combs, Jr.

Abstract: Evidence that selenium supplementation can reduce cancer risk is difficult to incorporate in nutrition thinking in which "nutritional essentiality" is the central concept. That concept, which defines nutrient need in terms of indispensability in diets and irreplaceable function in preventing specific deficiency disorders, was not developed to accommodate the function of a nutrient in reducing the risk to chronic disease, particularly when that function is not obligate but may be among several involved in maintaining good health. The findings of Clark et al. (JAMA 276, 1957–1963, 1996; Br J Urol 81, 730–734, 1998) suggest that selenium intakes of approximately twice the levels of the new Recommended Dietary Allowance or more can have such beneficial health effects. Because these intakes are above those required to support its accepted essential biochemical functions and because the maintenance of good general health as much as the prevention of specific deficiency disorders is the goal of public health, it is appropriate to reassess the nutritional essentiality paradigm. This discussion of the development and outcomes of the clinical intervention trials of Larry Clark and colleagues is presented in light of these issues.

Introduction

The reports by Larry Clark and colleagues (1,2) that selenium (Se) treatment could reduce cancer risks did more than extend a well-supported hypothesis into human clinical experimentation. They challenged contemporary thinking about the scope of functions of nutrients, and they are likely to change the very conceptualization of nutritional essentiality. It is the purpose of this article to address these impacts, which I know my friend Larry understood. To do so, it is useful to consider Se in its historical context.

Questioning Essentiality

Only 20 years ago, nutritionists were still debating whether Se was a nutrient "in its own right" or merely some sort of factor that "spared" the need for vitamin E and whether Se was required at all by humans. Such questions were, of course, reasonable in light of the knowledge then available. More than two decades of animal studies had consistently shown that Se deprivation produced lesions in smooth and skeletal muscles, capillaries, and/or liver; however, few, if any, effects were seen in the presence of adequate dietary vitamin E (3). Furthermore, there was no compelling evidence of adverse health effects associated with low Se status among well-studied human populations, most notably, Finland and New Zealand (4). Se was referred to as "a nutrient in search of a disease." This was the context of nutrition thinking when, in 1982, Larry Clark approached Bruce Turnbull and me about mounting a clinical intervention trial using a dietary supplement of Se.

This is not to suggest that Larry's idea was in any way ill-founded. In fact, by the early 1980s, three major research findings had generated huge interest in Se. Rotruck et al. (5) had shown that Se was an essential constituent of the antioxidative enzyme glutathione peroxidase (GPx). This finding, and the subsequent research it stimulated, allowed for the first time a mechanistic understanding of the well-known nutritional "sparing" of Se and vitamin E. Then, researchers in China (6) reported in the English language literature a cardiomyopathy (Keshan disease) that had an endemic distribution corresponding to that of severe Se deficiencies in soils, plants, and local livestock. Importantly, Keshan disease could be effectively prevented by Se supplementation. In addition, several groups had demonstrated that Se could, at least at high levels, reduce tumor yields in a variety of animal models (see reviews in Refs. 7–9). Furthermore, a body of supporting epidemiological and animal tumor model data had developed in the wake of the suggestion of Shamberger and Frost (10) that Se status was related to cancer risk. Larry was aware of these developments, and, through his doctoral research on the epidemiology of skin cancer, he knew that excess rates of basal and squamous cell cancers of the skin occurred along the lower eastern US seaboard, an area that Kubota et al. (11) had mapped among the lowest soil-Se areas of the country. Furthermore, Larry had found that nonmelanoma skin cancer risk in that region was associated with relatively low plasma Se concentrations,

The author is affiliated with the Division of Nutritional Sciences, Cornell University, Ithaca, NY 14853.

particularly in subjects with low plasma total carotenoids (12,13). Although the momentum of the field was moving toward mechanisms of action of Se and associations of the element with chronic diseases, Larry proposed to test what he called "the selenium hypothesis" in humans. He was the first person to do so.

So, when the first patients (with recent basal and/or squamous cell carcinomas of the skin) were recruited for what was to be called the Nutritional Prevention of Cancer (NPC) trial, the idea that supplemental dietary Se could reduce cancer risk was considered a plausible hypothesis but one supported only by animal studies. Furthermore, that hypothesis related to a trace element that, having been recognized as nutritionally important only in the late 1950s (14), was still poorly understood in terms of its nutrition and health implications.

Considering the Concept of Nutritional Essentiality

Consider that, for a century, thinking in the field of nutrition has been centered on the functions of those factors in the external chemical environments of organisms that are specifically required for normal physiological functions, including growth, reproductive success, survival, and freedom from clinical/metabolic disorders. Such factors, including vitamins, minerals, amino acids, fatty acids, and water, are called "essential" in the sense that each is indispensable from diets, and a diet is considered nutritionally "adequate" if it contains each of these essential nutrients at levels that meet or exceed known needs. Nutritional essentiality is the dominant concept of the field of nutrition. It emerged with the paradigm shift fostered by the studies of Eijkmann, Hopkins, Funk, Goldberger and others (15) that gave birth to the field with the recognition that health could be affected by diet quality. For a century, the science of nutrition developed under the nutritional essentiality paradigm, which holds that certain nutrients prevent ill health in very specific and irreplaceable ways. Nutrient deficiency diseases have played important roles in the development of nutrition knowledge: their specific prevention has been used to define nutrient essentiality and to establish dietary recommendations. Indeed, unless a clinical disease has been related specifically to the deprivation of a certain nutrient, then that nutrient has not been considered "essential."

How, then, was one to think about Se and human health in 1982? By that time, consensus had developed that the element should, indeed, be considered essential; its role in GPx (5) was sufficient to close that argument. In addition, Keshan disease, which is now suspected as having a viral etiology potentiated by antioxidant (e.g., Se-vitamin E) deficiency (16), was taken by many as a human Se-deficiency disease. Although there was no Recommended Dietary Allowance (RDA) of Se, in 1980 an estimated safe and adequate daily dietary intake was set: 50–200 µg/day (17). An RDA for the element would not come for another 9 years (18). Ultimately, an RDA was derived from a single study of the GPx

responses of young, Se-deficient Chinese men (19). That value was subsequently challenged (20), but the narrow and somewhat flawed database on which it was based is the same one that was used in the reformulation of the RDA as part of the dietary reference values (DRIs) (21). In all these cases, dietary recommendations for Se followed the nutritional essentiality paradigm and were designed to ensure adequate nutritional status as defined by optimal GPx expression. Implicit in this approach is that GPx is a relevant and sufficient indicator of the health-related function(s) of Se in the body. From this, it follows that supplemental Se can benefit health to the extent that it can correct residual, suboptimal Se status in a population.

Extending Nutritional Essentiality to Include the Reduction of Chronic Disease Risk

Americans, on average, appear to be adequately nourished, at least in the traditional sense, with respect to Se. For example, the subjects entering the NPC trial had, with very few exceptions, apparently adequate plasma Se levels (Fig. 1). Of the 1,312 subjects that were randomized in the trial, only 6 had initial plasma Se levels <80 ng/ml. This level may be taken as a criterion of nutritional adequacy, inasmuch as it corresponds to the amount of Se contained in maximally expressed plasma selenoproteins (22) as well as to the upper limit of GPx responses to Se supplements in healthy people (23). Although, in the face of these data, the possibility could not be excluded that apparently Se-adequate subjects may not have full selenoprotein expression in critical cells/tissues, a simpler interpretation was that any cancer-protective effects we might see would have to be explained on the basis of Se acting through nonselenoprotein mechanisms.

The latter hypothesis, which in those days was not widely held, was in fact strongly supported by the results of animal tumor model studies some hundred of which had been published by the mid-1980s. Two-thirds of those had shown Se to be antitumorigenic (half showing tumor yield reductions of >50%) when supplemented at supranutritional levels (e.g., 1–2.5 mg/kg) to diets that were apparently adequate

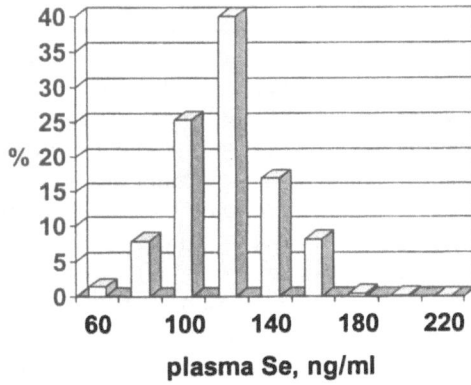

Figure 1. Distribution of plasma Se levels of Nutritional Prevention of Cancer (NPC) trial subjects on entry to trial. [Data are from unpublished technical reports of Clark and co-workers and from Combs et al. (35).]

(e.g., 0.1–0.5 mg/kg) in Se (7,8). Such responses, although highly relevant to considerations of health, are simply not addressed by the paradigm of nutritional essentiality, inasmuch as it is clear that cancer was not a specific, Se-deficiency disease. How, then, are nutritionists and others to consider the putative effects of a nutrient in reducing risk to a chronic disease, particularly when such effects may require high-level exposures?

The conventional answer to this question is to call such effects "pharmacological." Doing so, however, is ignoring the indication discussed below that cancer-protective levels of Se appear to be well within the normal range of Se intakes of Americans. Furthermore, it is not addressing three fundamental problems of the nutritional essentiality paradigm. The first concerns the connotation of a high degree of specificity in the nutrient-health linkage, which occurs in frank nutrient deficiency. Cannot a nutrient benefit health in a general but nonspecific way, such as the case of vitamin E and Se, for which nutritional "sparing" is well established? The second problem concerns the implicit focus on preventing such deficiencies instead of supporting optimal health. After all, frank single-nutrient deficiency disorders, particularly in industrialized societies, tend to be rare. In most public health contexts, the central concerns are for optimum health broadly conceptualized as a state of physical, mental, and emotional well-being and not merely freedom from disease or absence of an ailment. The third problem is that nutrient requirements can vary considerably, depending on the end point chosen as the defining criterion. These problems are not merely semantic; they go to the heart of several key questions in determining nutrient needs and adequacy.

The results of the NPC trial (1,2) should force the nutrition community to address these issues. Those results, which showed Se supplementation to reduce cancer risk, go to the general question of which level of nutrient need should define a dietary requirement. Should not such an outcome, clearly important to public health, be considered in formulating RDAs? It would seem obvious that it should. Thus it was both surprising and unfortunate that the panel that recently developed the DRIs for Se (21) gave only superficial consideration to what has become a voluminous literature on Se antitumorigenesis (see reviews in Refs. 9, 24, and 25). Citing Blot et al. (26) and Clark et al. (1), but not Yu et al. (27), the panel found those results "compatible with the possibility that intakes of Se above those needed to maximize selenoproteins have anticancer effects in humans" and, without considering their implications, wrote that "they do not, however, allow conclusions concerning dietary Se requirements" (21). In making that choice, the DRI panel bypassed the issue of cancer outcome relevance to RDA considerations, thus leaving it open.

Therefore, the following question remains: How should nutritional guidance be given in cases where nontraditional intakes of nutrients can reduce risk to chronic disease in nonspecific ways? The NPC trial is a case study for this question. It found that risks to cancer mortality and the incidences of total carcinomas as well as cancers of the lung, prostate, and colon-rectum each appeared to be reduced in response to Se supplementation at 200 µg/day to normal diets that appeared to have provided an additional ≥85 µg Se/day. This level of total Se intake is several times greater than that necessary to support the maximal expression of the known selenoenzymes. Although the DRI panel offered Americans guidance to keep their selenoproteins well expressed (i.e., with an RDA of 55 µg of Se), who will not be far more interested in the prospect of reducing his/her cancer risk?

Considering Efficacy

In considering the incorporation of cancer risk reduction in the setting of dietary recommendations for Se, the issues of dose and safety must be addressed. The optimal dose of an oral form of Se would be the minimal amount that would both prevent deficiency and reduce cancer risk, doing so with minimal risks of adverse effects. Phrased in metabolic terms, the optimal dose would be the minimal amount of bioavailable Se that would not only support optimal expression of selenocysteine enzymes but also support the production of tumorigenic Se metabolites while not producing toxicity. The studies of Ip and colleagues (28–33) and Ganther (34) have pointed to hydrogen selenide (H_2Se), methylselenol [$(CH_3)SeH$], and perhaps other reduced, methylated Se metabolites as mediators of Se-antitumorigenesis, but analytic methods are not available to determine these species in accessible tissues. The metabolic production of these putatively antitumorigenic Se metabolites, therefore, must be estimated from other parameters of Se status.

The best data for making these sorts of estimations come from the NPC trial (1,2,37; unpublished technical reports). Although those data show that Se supplementation in addition to a background dietary intake of ≥85 µg Se/day was effective in reducing risk of each of the most prevalent cancers in that cohort, they also show that the protective effect was greatest for subjects who entered the trial with plasma Se levels in the lower tertiles of the cohort. Specifically, subjects entering the trial with plasma Se levels <106 ng/ml showed not only the highest rates of subsequent cancer but also the strongest apparent protective effects of Se supplementation (Figs. 2 and 3). The tertile of subjects entering with plasma Se levels >121 ng/ml showed no cancer-protective benefits from taking the Se supplement. In other words, the NPC trial results suggest that the normal Se intakes of at least one-third of the subjects were sufficient to reduce their cancer risks and that Se supplementation of the other two-thirds of the cohort rendered the observed cancer-protective benefits. This would suggest that the plasma level of ~120 ng/ml (i.e., ~1.5 nmol/ml) may be optimal for cancer protection and that strategies to reduce cancer risk using a selenomethionine-rich supplement such as Se-enriched yeast might only need to raise Se status above that level. The extensive data of Yang et al. (36), which cover three orders of magnitude for each variable, are the best available for es-

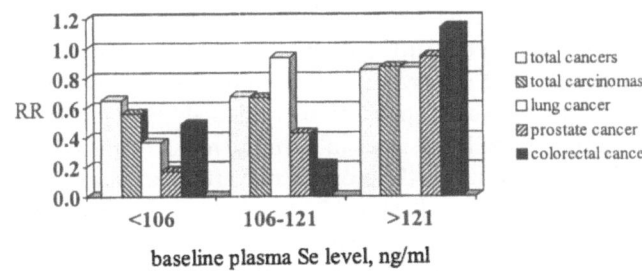

Figure 2. Protective effect of supplemental Se (200 µg/day) in NPC trial was greatest for subjects entering trial in lower tertiles of plasma Se. [Data are from unpublished technical reports of Clark et al., Clark et al. (2), and Combs et al. (35).] Ratio of cancer rates between Se-treated and placebo groups is presented as relative risk (RR).

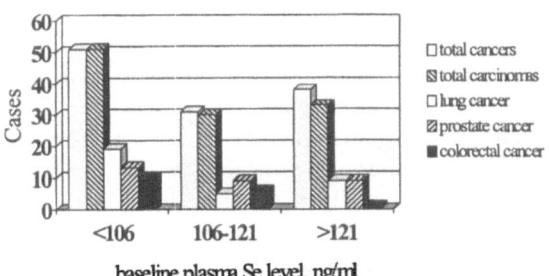

Figure 3. Cancer incidence among placebo-treated subjects in NPC trial by their plasma Se level on entry to trial. [Data are from unpublished technical reports of Clark et al., Clark et al. (2), and Combs et al. (35).]

timating the daily total Se intake associated with a given plasma Se level. They show that the relationship can be described by a linear function: $\log y = 1.623 \log x + 3.455$, where y is Se intake (in µg/day) and x is plasma Se (in mg/l). This equation would predict that an intake of ~91 µg Se/day would support a plasma Se level of 120 ng/ml; however, in using this equation previously (37), we believed that it underestimated, perhaps by as much as one-third, the dietary Se intakes of NPC trial subjects, apparently because of differences in the average body weights of the subjects between the studies: estimated to be 60 kg in the study of Yang et al.

and 77 kg in the NPC trial. Correcting for this difference and assuming similar fractional intakes of selenomethionine and related compounds in the foods consumed in the two studies, it would appear that dietary Se intakes of ≥96 µg/day (women) and 120 µg/day (men) are required to support the plasma Se concentrations at the 120 µg/ml level for Americans. With the assumption of linearity in this dose response, that intake would correspond to ~1.5 µg Se/kg body wt/day. These levels are 175% and 218%, respectively, of the revised RDA (21).

Considering Safety

It is clear that such levels are safe. After investigating cases of selenosis of dietary origin in rural China, Yang et al. (35) concluded that "the minimal blood Se concentration of 1.054 mg/dl (i.e., a plasma level of ca. 855 ng/ml) would have to be taken as approximately the marginal level of Se-toxicity," on which basis the whole blood level of 1 mg Se/dl (i.e., 810 ng Se/ml plasma) was taken as the upper safe limit for most healthy adults in deriving the reference dose for Se (38). Such levels would require regular intakes on the order of 2 mg/day.

Clark and colleagues (37; unpublished technical report) found no evidence of adverse effects of Se supplementation at 200 or 400 µg/day. After some 8,271 person-yr of observation in 1983–1993, only 203 of 1,312 NPC trial subjects had voluntarily withdrawn from the study (Table 1). That number was distributed in the placebo (158) and Se-treatment (145) groups and included only 39 subjects (14 in the placebo group and 25 in the Se group) complaining of "side effects." Of these 39 subjects, only a few (5 in each treatment group) complained of changes consistent with selenosis (brittle hair, white-patched/brittle nails, skin rash, "garlic" breath). The higher level of Se supplementation was used in a second intervention trial (the "400-µg trial") that involved 424 subjects at a single clinic (in Macon, GA) ran-

Table 1. Safety Indicators in NPC and "400-µg" Trials

		Withdrawals			High Plasma Se Levels			
	n	All causes	Side effects	Dermatological complaints[a]	>350 ng/ml[b,c]	High 5 samples	>500 ng/ml[b,c]	High 5 samples
NPC trial[d]								
Placebo[e]	659	158	14	5	0			
Se (200 µg/day)[f]	653	148	25	5	20 (12)	721,[g] 609, 424, 409, 408		
400-µg trial[h]								
Placebo[e]	212	42	10	3			0	
Se (400 µg/day)[f]	212	32	7	0			8 (5)	564, 556, 542, 539, 524

a: Dermatological complaints included brittle hair and/or nails, white-patched nails, and skin rash (no garlic breath was reported).

b: Respective trial safety monitoring level.

c: Values in parentheses represent number of total samples.

d: Data are from Nutritional Prevention of Cancer (NPC) trial (1).

e: Bakers' yeast.

f: Se-enriched bakers' yeast.

g: Subjects reported regular consumption of Brasil nuts.

h: Data are from unpublished technical reports of Clark et al. (1994).

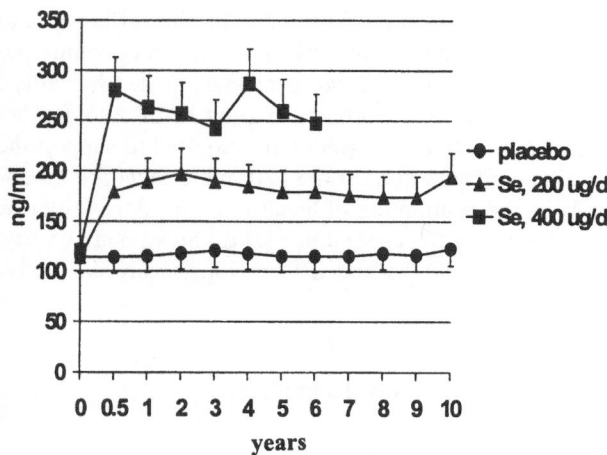

Figure 4. Plasma Se responses to dietary supplements of Se (200 and 400 µg/day) as Se-enriched yeast. [Data are from unpublished technical reports of Clark et al. and Clark et al. (1).]

domized to the bakers' yeast placebo or 400 µg Se/day as Se-enriched yeast. The results showed that the 400 µg Se/day supplement increased plasma Se levels to ~250 ng/ml, i.e., ~32% greater than the levels produced by the 200 µg Se/day supplement in the NPC trial (Fig. 4). After 1,220 person-yr of observation, only 74 subjects from the 400-µg trial had withdrawn (42 in the placebo group and 32 in the Se group), including 17 complaining of side effects (10 in the placebo group and 7 in the Se group; Table 1). Of these 17 subjects, only 3 (in the placebo group) complained of changes consistent with selenosis (e.g., changes in hair, nails, or skin or garlic breath). These subjects underwent regular biannual clinical chemical evaluation, which revealed no significant changes except for slight elevations in serum alanine aminotransferase activities in seven subjects (2 in the placebo group and 5 in the Se group). All but one of these elevations were transient. In the 400-µg and NPC trials, the body weights of the subjects with higher plasma Se levels tended to be lower than the average body weight, but in neither trial did subjects complaining of side effects show plasma Se levels that were remarkable in comparison with other subjects in their respective treatment group. In no case did a subject complain of garlic breath or show plasma Se levels near the US Environmental Protection Agency upper safe limit.

Changing a Paradigm

Current evidence would suggest that Se can be antitumorigenic in two possible ways: 1) as an essential nutrient in the classical sense by providing the catalytic centers of antioxidant enzymes and 2) as a source of metabolites that directly inhibit tumorigenesis. According to this view, Se is seen as potentially affecting a number of anticarcinogenic activities depending, in part, on the form and level of the Se dose. To the extent that these effects involve enhancements in the expression of selenoproteins [e.g., in Se supplementation of individuals with Se intakes less than ~40 µg/day (18)], they would appear to be most relevant to the support

of antioxidant defense and the inhibition of tumor initiation. To the extent that they involve supranutritional Se exposures [perhaps 96–300 µg/day, as suggested by the NPC trial results (1)], they would appear to be mediated by Se metabolites active in the later stages of tumorigenesis. It was Larry Clark's hypothesis that the proapoptotic effects of Se may be most important in cells that do not have p53 mutations, which inhibit programmed death of transformed cells. If this is so, and because p53 expression is a late-stage phenomenon in many cancers, Se compounds might be expected to be effective against initiated and transformed cells and, thus, to have enormous public health significance.

Evidence (reviewed in Refs. 24, 25, and 33) suggests that antitumorigenic activities can be attributed to metabolites of several naturally occurring forms of the element: the common dietary selenoamino acids selenomethionine and selenocysteine, methylated Se compounds such as methylselenocysteine, and the inorganic Se salts selenite and selenate. With varying efficiencies, these species can be converted to a number of Se metabolites. If CH_3SeH is, indeed, the key antitumorigenic metabolite in humans, then cancer-prevention strategies should not simply target apparently effective plasma levels of total Se (e.g., 120 ng/ml, as suggested by the NPC trial results); they should target the optimal production of that metabolite.

As important as understanding the mechanisms underlying these protective effects is the translation of such understanding into forms useful for the consuming public as well as public health policy makers. To that end, the concept that a nutrient can effect such nontraditional health outcomes as cancer risk must be addressed directly. Public interests may not be best served unless chronic disease prevention is carefully considered in developing RDAs. The failure to do so, including the labeling of such effects as pharmacological, is to limit the implementation of this aspect of diet-health knowledge to the development of specialty products instead of encouraging its broad implementation throughout the food system. Such a perspective is necessitated by findings such as those from the NPC trial. Because it addresses the contemporary need not only to prevent nutrient deficiencies but also to support good general health, such a "healthy diet" perspective is rapidly becoming the new paradigm for nutrition. The development of this paradigm will call for the resolution of many current issues, and results such as those of the NPC trial (1) are certain to be at the vortex of those discussions.

Acknowledgments and Notes

Address correspondence to Dr. G. F. Combs, Jr., 122 Savage Hall, Cornell University, Ithaca, NY 14853. Phone: (607) 255-2140. FAX: (607) 255-1033. E-mail: gfc2@cornell.edu.

Submitted 5 December 2000; accepted in final form 4 January 2001.

References

1. Clark LC, Combs GF Jr, Turnbull BW, Slate EH, Alberts D, et al.: The nutritional prevention of cancer with selenium 1983–1993: a randomized clinical trial. *JAMA* **276,** 1957–1963, 1996.

2. Clark LC, Dalkin B, Krongrad A, Combs GF Jr, Turnbull BW, et al.: Decreased incidence of prostate cancer with selenium supplementation: results of a double-blind cancer prevention trial. *Br J Urol* **81**, 730–734, 1998.

3. Combs GF Jr and Combs SB: Selenium deficiency diseases of animals. In *The Role of Selenium in Nutrition*. New York: Academic, 1986, pp 266–326.

4. Combs GF Jr and Combs SB: Selenium in human nutrition and health. In *The Role of Selenium in Nutrition*. New York: Academic, 1986, pp 327–400.

5. Rotruck JT, Ganther HE, Swanson AB, Hafeman DG, and Hoekstra WG: Selenium: biochemical role as a component of glutathione peroxidase. *Science* **179**, 588–590, 1972.

6. Yang GQ, Wang GY, and Yin TA: Relation between the distribution of Keshan disease and selenium status. *Acta Nutr Sin* **4**, 191–200, 1982.

7. Combs GF Jr and Combs SB: Selenium and cancer. In *The Role of Selenium in Nutrition*. New York: Academic, 1986, pp 413–462.

8. Combs GF Jr: Selenium. In *Nutrition and Cancer Prevention*, T Moon and M Micozzi (eds). New York: Dekker, 1989, pp 389–420.

9. El-Bayoumey K: The role of selenium in cancer prevention. In *Practice of Oncology*, 4th ed, DeVita VT, Hellman S, and Rosenberg SS (eds). Philadelphia, PA: Lippincott, 1991, pp 1–15.

10. Shamburger RJ and Frost DV: Possible protective effect of selenium against human cancer. *Can Med Assoc J* **104**, 82–89, 1969.

11. Kubota J, Allaway WH, Carter DL, Cary EE, and Lazar VA: Selenium in crops in the United States in relation to selenium-responsive diseases of animals. *Agric Food Chem* **15**, 488–495, 1967.

12. Clark LC, Graham GF, Crounse R, Hulka BS, and Shy CM: Plasma selenium and skin cancer: a case control study. *Nutr Cancer* **6**, 13–21, 1984.

13. Clark LC, Graham GF, Bray J, Turnbull BW, Hulka BS, et al.: Nonmelanoma skin cancer and plasma selenium: a prospective cohort study. In *Selenium in Biology and Medicine*, Combs GF Jr, Spallholz JE, Levander OA, and Oldfield JE (eds). Eastport, CT: AVI, 1984, pt B, pp 1122–1134.

14. Schwartz K and Foltz CM: Selenium as an integral part of factor 3 against dietary necrotic liver degeneration. *J Am Chem Soc* **79**, 3292–3293, 1957.

15. Combs GF Jr: Discovery of the vitamins. In *The Vitamins: Fundamental Roles in Nutrition and Health*, 2nd rev ed. New York: Academic, 1998, pp 10–43.

16. Beck MA: Rapid genomic evolution of a non-virulent Coxsackievirus B_3 in selenium-deficient mice. *Biomed Environ Sci* **10**, 307–315, 1997.

17. Committee on Dietary Allowances: *Recommended Dietary Allowances*, 9th rev ed. Washington, DC: National Academy Press, 1980.

18. Subcommittee on the Tenth Edition of the Recommended Dietary Allowances: *Recommended Dietary Allowances*, 10th ed. Washington, DC: National Academy Press, 1985.

19. Yang GQ, Qian PC, Zhu LZ, Huang JH, Liu SJ, et al.: Human selenium requirements in China. In *Selenium in Biology and Medicine*, Combs GF Jr, Spallholz JE, Levander OA, and Oldfield JE (eds). Westport, CT: AVI, 1987, vol A, pp 589–594.

20. Combs GF Jr: Essentiality and toxicity of selenium: a critique of the recommended dietary allowance and the reference dose. In *Risk Assessment of Essential Elements*, Mertz W, Abernathy CO, and Olin SS (eds). Washington, DC: ILSI, 1994, pp 167–183.

21. Panel on Dietary Antioxidants and Related Compounds: *Dietary Reference Intakes for Vitamin C, Vitamin E, Selenium and β-Carotene and Other Carotenoids*. Washington, DC: National Academy Press, 2000.

22. Hill KE, Xia Y, Åkesson B, Boeglin ME, and Burk RF: Selenoprotein P concentration in plasma as an index of selenium status in selenium-deficient and selenium-supplemented Chinese subjects. *J Nutr* **126**, 138–145, 1996.

23. Nève J: Human selenium supplementation as assessed by changes in blood selenium concentration and glutathione peroxidase activity. *J Trace Elem Med Biol* **9**, 65–73, 1995.

24. Combs GF Jr and Gray WP: Chemopreventive agents: selenium. *Pharmacol Ther* **79**, 179–192, 1998.

25. Combs GF Jr and Clark LC: Selenium and cancer. In *Nutritional Oncology*, Heber D, Blackburn GL, and Go VLW (eds). New York: Academic, 1999, pp 215–222.

26. Blot WJ, Li JY, Taylor PR, Guo W, Dawsey S, et al.: Nutrition intervention trials in Linxian, China: supplementation with specific vitamin/mineral combinations, cancer incidence, and disease-specific mortality in the general population. *JNCI* **85**, 1483–1490, 1993.

27. Yu SY, Zhu YJ, and Li WG: Protective role of selenium against hepatitis B virus and primary liver cancer in Qidong. *Biol Trace Elem Res* **56**, 117–124, 1997.

28. Ip C: Lessons from basic research in selenium and cancer prevention. *J Nutr* **128**, 1845–1854, 1998.

29. Ip C, Hayes C, Budnick RM, and Ganther HE: Chemical form of selenium, critical metabolites, and cancer prevention. *Cancer Res* **51**, 595–600, 1991.

30. Ip C, El-Bayoumy K, Upadhyaya P, Ganther H, Vadanavikit S, et al.: Comparative effect on inorganic and organic selenocyanate derivatives in mammary cancer prevention. *Carcinogenesis* **15**, 187–192, 1994.

31. Ip C and Ganther H: Activity of methylated forms of selenium in cancer prevention. *Cancer Res* **50**, 1206–1211, 1990.

32. Ip C and Ganther H: Novel Strategies in Selenium Cancer Chemoprevention Research. In *Selenium in Biology and Human Health*, Burk RF (ed). New York: Springer-Verlag, 1993, pp 170–180.

33. Lü J, Jiang C, Kaeck M, Ganther H, Vadhanavikit S, et al.: Dissociation of the genotoxic and growth inhibitory effects of selenium. *Biochem Pharmacol* **50**, 213–219, 1995.

34. Ganther HE: Selenium metabolism, selenoproteins and mechanisms of cancer prevention: complexities with thioredoxin reductase. *Carcinogenesis* **20**, 1657–1666, 1999.

35. Combs GF Jr, Clark LC, and Turnbull BW: An analysis of selenium and cancer prevention. In *Proceedings of the Sixth International Symposium on Selenium in Biology and Medicine*. Padua, Italy: University of Padua. In Press.

36. Yang GQ, Yin S, Zhou R, Gu L, Yan B, et al.: Studies of safe maximal daily dietary Se intake in a seleniferous area in China. II. Relation between Se intake and the manifestations of clinical signs and certain biochemical alterations in blood and urine. *J Trace Elem Electrolytes Health Dis* **3**, 123–130, 1989.

37. Combs GF Jr, Clark LC, and Turnbull BW: Responses to selenium supplementation in Americans. In *Proceedings of the Fifth International Symposium on Selenium in Biology and Medicine*. Nashville, TN: Vanderbilt University, 1992, p 126.

38. Poirier KA: Summary of the derivation of the reference dose for selenium. In *Risk Assessment of Essential Elements*, Mertz W, Abernathy CO, and Olin SS (eds). Washington, DC: ILSI, 1994, pp 157–166.

NUTRITION AND CANCER, *40*(1), 12–17

Se-Methylselenocysteine: A New Compound for Chemoprevention of Breast Cancer

D. Medina, H. Thompson, H. Ganther, and C. Ip

Abstract: Selenium compounds have attracted renewed interest as chemopreventive agents for human cancer on the basis of the pioneering intervention study by Clark and coworkers. The rodent mammary gland has been used extensively as a model for examining the chemopreventive activities of inorganic and organic selenium compounds. This review summarizes the rationale and results for use of a new organic selenium compound, Se-methylselenocysteine, which exhibits greater efficacy as a chemopreventive agent than several previously used selenium compounds in experimental models of breast cancer and has potential for use in human populations.

Introduction

The intervention study by Clark et al. (1) demonstrated that a modest supplement of selenized yeast had a striking effect on cancer morbidity and mortality. The morbidity and mortality due to prostate, colorectal, and lung cancer decreased by ≥50% in individuals given selenium at 200 µg/day for a mean duration of 4.5 yr. This landmark study confirmed the existing evidence in the literature that suggested that selenium was a protective agent for human cancer. The earlier epidemiological studies were correlational studies relating selenium levels in crops with mortality rates (2,3) and case-control and prospective studies demonstrating increased cancer risk with low blood selenium levels (4,5).

Among the numerous organs in which tumorigenesis is affected by selenium status, the mammary gland ranks as the most intensively studied over the past 20 years (6). Selenium compounds are among the few agents that inhibit multiple mammary tumor models, which include mouse mammary tumor virus–induced mouse mammary tumorigenesis, chemical carcinogen–induced mammary tumorigenesis in the mouse and rat, and spontaneous tumorigenesis in the mouse (7,8). The early studies in mouse and rat models demonstrated that a nontoxic dose of ~3–5 ppm sodium selenite resulted in ≥50% inhibition of mammary tumorigenesis. Se-

lenium exhibited a marked stage specificity, in that early-stage (i.e., preneoplastic stages), but not normal, mammary development or existing mammary tumor growth was affected by supplemental selenium status (8,9). These early studies led to a concerted effort on our part in the last decade to determine the mechanism of selenium inhibition of mammary tumorigenesis and to develop new selenium compounds. This review focuses on one agent, Se-methylselenocysteine, which has greater and more specific activity than earlier selenium-containing agents on experimental mammary tumorigenesis that we have tested over the past 15 years.

The Guiding Principle

Ip and Ganther (10,11) proposed a hypothesis, based on a consideration of selenium metabolism (Fig. 1) that selenium compounds that directly enter the methylated pool were more effective chemopreventive agents than selenium compounds that were metabolized through the H_2Se pool. Experiments examining the chemopreventive activities of selenobetaine and Se-methylselenocysteine (which generate monomethylated selenium) compared with Na_2SeO_3 and selenomethionine (which are metabolized to H_2Se) supported the basic hypothesis (11,12). Selenobetaine and Se-methylselenocysteine were more efficacious than the other two forms as chemopreventive compounds in the carcinogen-induced rat mammary tumorigenesis model. This hypothesis has gained further support from recent experiments that methylseleninic acid, a direct precursor of methylselenol, acts much more rapidly than Se-methylselenocysteine in inhibition of DNA synthesis in in vitro mammary cell lines (13,14). Table 1 shows the relative chemopreventive activity and tolerance index of six compounds.

An extension of this experiment was an examination of the effects of the carbon side chain attached to selenocysteine. The results indicated that the length of the side chain was proportional to the chemopreventive activity. The Se-allylselenocysteine reduced tumor yield by 86% compared

D. Medina is affiliated with the Department of Molecular and Cellular Biology, Baylor College of Medicine, Houston, TX 77030. H. Thompson is affiliated with the AMC Cancer Research Center, Denver, CO 80214. H. Ganther is affiliated with the Department of Nutritional Sciences, University of Wisconsin, Madison, WI 53706. C. Ip is affiliated with the Roswell Park Memorial Institute, Buffalo, NY 14263.

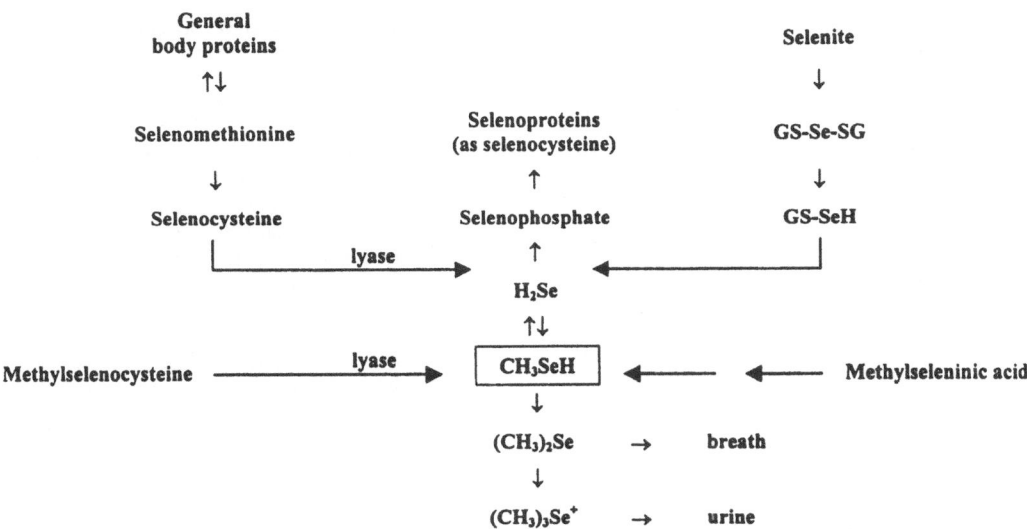

Figure 1. General selenium metabolic pathway. Selenomethionine is incorporated into proteins or converted to selenocysteine. Selenocysteine is metabolized to H_2Se by selenocysteine lyase. Se-methylselenocysteine (and other alkylselenocysteines or monomethylated selenium-generating compounds) is metabolized directly to methylselenol by a cysteine conjugate, β-lyase. Selenite is metabolized to H_2Se, which is a precursor for generating selenium for synthesis of selenoproteins. H_2Se is eliminated via progressive methylation to methylselenol, dimethylselenide, and trimethylselenonium. Methylseleninic acid is a direct precursor to methylselenol.

Table 1. Anticarcinogenic Efficacy of Different Selenium Compounds in the Chemically Induced Mammary Cancer Model

Compound	Dose of Se for 50% Inhibition, ppm	MTD,[a] ppm
Se-allylselenocysteine	≤1	ND
Se-methylselenocysteine	2	5
Selenobetaine	2	5
Methylseleninic acid	2	5
Selenite	3	4
Selenomethionine	4–5	5–6

a: Maximum tolerable dose (MTD) is defined as the dose that produces the first indication of a significant suppression in body weight gain (usually 7–10%, $p < 0.05$), most likely as a result of food avoidance. ND, not done.

with a 50% reduction by the propyl- or Se-methylselenocysteine. Tissue selenium levels did not correlate with differences in chemopreventive activity. The studies also demonstrated the presence of a β-lyase capable of cleaving the Se-alkyl group from the amino acid. These results suggest that compounds even more effective than Se-methylselenocysteine are feasible and demonstrate a useful chemoprevention strategy of administering selenium in the form of a stable, alkylated selenoamino acid that releases an active selenium compound through lyase action (15,16). Furthermore, it is likely that the Se-alkyl moiety may also be a determinant of the biological effect.

In summary, the in vivo tumor experiments have confirmed the validity of the hypothesis originally formulated by Ganther and Ip and have led to the synthesis and characterization of a novel selenium chemopreventive agent (alkylselenocysteine, e.g., Se-methylselenocysteine) that is more efficacious than traditional forms of selenomethionine

and sodium selenite. Se-methylselenocysteine occurs naturally in plants; in fact, it is present at high levels in the selenium accumulator plants of the genus *Astragalus*. Se-methylselenocysteine is readily converted to the methylated metabolite in mammalian cells via β-lyase (15) (Fig. 1). It is the major form of selenium in selenium-enriched garlic (17), a food form of garlic that exhibits a high chemopreventive activity due to its selenium content.

Mechanism of Action

Selenium compounds have been reported to affect numerous cellular processes and molecular pathways. Some of the molecular targets recently identified as modulated by selenium include selenoenzymes (e.g., redox enzymes) (18, 19), protein kinase C (20–23), protein phosphorylation (21, 24), phase I and II enzyme activities (25,26), nuclear factor-κB activity (27), specific kinases (28), *gadd* activation (29), tumor suppressor gene p53 (30), activator protein-1 binding to DNA (31), DNA cytosine methyltransferase activity (32), and eicosanoid biosynthesis (33). The hypothesis advanced by Ip and Ganther, that selenium compounds which directly enter the methylated pool would be more effective chemopreventive agents, gave rise to an additional line of investigation. Thompson and associates (34–38) posed the following question: Can the toxicological effects of selenium be dissociated from the effects that account for selenium anticancer activity? The ability to directly address this question was made possible by the availability of Se-methylselenocysteine and related methylselenol-generating compounds. In a series of experiments (34–38), the results of which are summarized in Table 2, it was determined that inorganic forms of selenium such as selenite that underwent

Table 2. In Vitro Effects of Selenium Compounds That Differ in Their Intracellular Metabolism[a]

Biochemical/Cellular Event[b]	Selenium Species That Undergo Reductive Metabolism[c]	Selenium Species That Enter Monomethylated Pool[d]	Comments
Membrane leakage	↑	Minimal changes	Correlated with DNA SSBs and necrosis
DNA SSB	↑	Not observed	Correlated with membrane leakage and necrosis
DNA DSB	↑	↑	Correlated with induction of apoptosis
Cell cycle arrest	S-G_2/M	G_1, S-G_2/M	G_1 block may result from specific effects on cell cycle regulatory molecules. S-G_2/M can be induced by alterations of cell cycle regulatory molecules or SSB.
Cell proliferation			
DNA synthesis	↓	↓	Measured as [^{3}H]thymidine incorporation
Cell doubling rate	↓	↓	Rate determined during a 24-h exposure period
Long-term proliferative potential	↓	↓	Measured as colony formation of treated cells in soft agar
Cell death			
Necrosis	Predominant	Limited	SSB observed in cells undergoing necrosis; cells with SSB can also be induced to undergo apoptosis
Apoptosis	Secondary	Predominant	DSB observed in cells undergoing apoptosis

a: Abbreviations are as follows: SSB, single-strand breakage; DSB, double-strand breakage.
b: Effects were observed in mouse mammary epithelial hyperplastic and tumor cell lines and in a mouse leukemia cell line.
c: Example of compounds of this type is sodium selenite.
d: Example of compounds of this type is Se-methylselenocysteine.

metabolism to H_2Se displayed a biological spectrum of activity that was distinct from forms of selenium such as Se-methylselenocysteine that directly entered the methylated pool (Fig. 1). Specifically, selenite induced a sequelae of effects that included inhibition of DNA synthesis, decreased cell doubling rate, and reduced long-term growth potential as well as the induction of cell death. However, these effects were accompanied by damage to cellular DNA and loss of cell membrane integrity. On the other hand, Se-methylselenocysteine and related compounds decreased DNA synthesis, cell doubling rate, long-term growth potential, and cell death, but in the absence of changes in membrane integrity or cellular DNA integrity measured as single-strand breaks in the DNA. Further investigations revealed distinctive differences in the mechanism of cell death induction by selenite vs. Se-methylselenocysteine. Although necrotic and apoptotic cells could be detected in cultures treated with either compound, the dominant type of death mediated by selenite was necrosis, whereas apoptosis was the predominant type of cell death mediated by Se-methylselenocysteine. These observations provided a context in which to distinguish between effects of some selenium compounds that appear to be nonspecific yet capable of inducing cell growth inhibition and cell death and those that may be specific to mechanisms of chemoprevention. Thus the two main cellular events that seem to be mediated by selenium and that are related to cancer prevention are inhibition of cell proliferation and induction of apoptosis.

A large number of experiments demonstrate that selenium compounds inhibit cell proliferation (6). Se-methylselenocysteine inhibits cell proliferation in multiple mouse mammary cell lines (37,39). The use of synchronized mouse mammary cells has shown that Se-methylselenocysteine blocks cells in the S and G_1 phases of the cell cycle. The block in the S phase is associated with a decrease in Cdk2

kinase activity and Cdk2 phosphorylation (14,21,39) and is fully reversible. Apparently, Se-methylselenocysteine does not affect the level and complex formation of Cdk2 inhibitors, such as p21 (Sinha and Medina, unpublished observations). The block of a specific cell cycle phase is probably dependent on a critical level of the ultimate selenium reactive form, i.e., methylselenol. Using synchronized cells in vitro, Se-methylselenocysteine generates methylselenol slowly; thus cells attain a critical level of this molecule 6 h after addition of the selenium compound, a time when the cells have reached the S phase. In contrast, methylseleninic acid generates methylselenol rapidly; thus a critical level of this molecule is attained soon after the cells are released from their growth arrest state, a time when the cells are still in the G_1 phase.

A more dramatic effect is observed on apoptosis. Thompson and Sinha (14,34,35,37,39) demonstrated that apoptosis is frequent in cells treated with Se-methylselenocysteine, as well as other selenium compounds. Apoptosis can be detected by DNA fragmentation, flow cytometry, and changes in function of apoptosis-related proteins. The first detectable molecular signs of apoptosis occur ≥8–12 h after the peak time period of inhibition of DNA synthesis.

These results raise important questions. First, what molecular pathway is mediated by selenium to induce apoptosis? There are multiple pathways, most of which converge on activated caspase-3 to activate apoptosis. Second, what is the upstream target of Se-methylselenocysteine? Se-methylselenocysteine has been shown to inhibit protein kinase C activity as a very early event, but whether this inhibition is related to subsequent alterations in Cdk2 or an apoptosis-related molecule is not known.

The use of in vitro cell lines provides important clues to the mechanism of selenium-mediated inhibition of cell growth; however, mammary cell growth in situ is affected by cell and tissue interactions as well as complex growth factor interactions that are difficult to replicate in a two-dimensional in vitro system. It has been known for several years that selenium compounds do not affect proliferation of normal mammary cells (or normal intestinal cells) in situ. Therefore, studies examining modulation of cell proliferation and cell cycle biomarkers in normal and transformed cells in an in situ setting are required to understand the mechanisms of selenium chemoprevention. A recent study by Ip et al. (40) addressed this issue. Normal mammary cells (i.e., terminal end buds and alveolar cells) and premalignant intraductal proliferating (IDP) cells were examined by immunohistochemical staining at 50 days of age and 6 wk after carcinogen treatment, respectively. Se-methylselenocysteine (3 ppm) did not alter bromodeoxyuridine labeling or proliferating cell nuclear antigen, cyclin D1, or p27^{Kip1} labeling in the terminal end buds or alveolar cells; however, the number of IDP lesions was reduced by 60%. The reduction in IDP lesions was not accompanied by decreases in bromodeoxyuridine labeling or the frequency of IDP cells expressing proliferating cell nuclear antigen or cyclin D1. Interestingly, the percentage of IDP cells expressing p27^{Kip1} increased slightly, but significantly. The above results are important, because they confirm the remarkable stage specificity of selenium compounds (6,8) and they suggest that Se-methylselenocysteine blocks carcinogenesis by a pathway that does not involve proliferation per se. An extension of the in vitro experiments would suggest that the apoptotic pathway is a logical candidate as the primary pathway affected by Se-methylselenocysteine in the whole tissue. Preliminary results indicate that apoptosis frequency is increased in IDP cells, but not in normal mammary cells, in Se-methylselenocysteine-treated rats (C. Ip, unpublished observations).

Other Possible Mechanisms

A recent provocative and exciting result on selenium chemoprevention was the demonstration that Se-methylselenocysteine leads to a significant reduction in angiogenesis (41). The decrease in angiogenesis was detected by intratumoral microvessel density and vascular endothelial growth factor in mammary carcinomas. Furthermore, methylseleninic acid, a precursor of methylselenol, decreased gelatinolytic activities of matrix metalloproteinase-2 in a human umbilical vein endothelial cell assay. The mechanisms of selenium regulation of vascular endothelial growth factor expression are speculative at this time and could involve redox modification of transcriptional regulators (42). The results are potentially important, because they emphasize selenium effects on the nontumor cell compartment of the tumorigenic process, e.g., the supporting stroma. If validated in future experiments, the results would imply a protective effect of selenium compounds on metastatic growth. Inasmuch as metastasis is not readily measured in the traditional rat mammary tumorigenesis models, there has been no suggestion as to a role of selenium in regulation of mammary metastasis.

Additionally, the results could partially explain the seeming paradox that selenium compounds do not affect proliferation and differentiation of normal mammary cells in vivo or the proliferation of growing tumor cells in vivo. Indeed, removal of selenium supplementation results in reexpression of tumor growth. The results would imply that stage specificity of selenium compounds might be due to their effects on a specific stage of preneoplastic progression, i.e., the acquisition of angiogenesis.

Selenoproteins and Selenium-Modified Proteins

Selenium interacts with cellular proteins in two distinct ways. The traditional selenoproteins incorporate selenocysteine in stoichiometric amounts as an integral amino acid in the structure of the proteins. The functions of selenium as an essential nutrient are believed to occur via these selenoproteins, such as glutathione peroxidases, iodothyronine deiodinases, and thioredoxin reductase (18,19). There is little evidence that the chemopreventive activities of selenium involve the traditional selenoproteins. In addition to these selenoproteins, experiments using Se-radioisotopes have demonstrated a few proteins that avidly bind selenium in nonstoichiometric amounts (43,44). Ganther (42) recently summarized arguments that cysteine clusters in proteins would be targets for binding by selenium, and subsequent modification by selenium formation of protein-Se adducts might affect various protein functions, such as redox signaling, catalytic activity, or receptor binding. This intriguing hypothesis is attractive, because it does not require a 1:1 stoichiometric ratio and finds support in recent results that demonstrate that protein kinase C catalytic function is inhibited by low levels of selenite (22,23,42). This hypothesis warrants further experimental verification. If validated, it would provide a biochemical explanation for some of the anomalous effects of selenium on cell function.

Future Studies

The future of Se-methylselenocysteine focuses on two areas. First, it is clear that more studies are necessary to understand the mechanism of action of this agent. The molecular targets for apoptosis and/or angiogenesis need more intensive examination to more precisely define the mechanisms of selenium inhibition of tumorigenesis and to provide possible molecular markers as surrogate end points. Second, Se-methylselenocysteine (and possibly Se-allylselenocysteine) holds promise as a true second-generation selenium chemoprevention compound because of its superior in vivo efficacy, virtually nonexistent toxicity, low body accumulation, and simple formulation. Phase I trials need to be pursued expeditiously so phase II intervention trials can be performed for several organ sites.

Summary

This short review of a new selenium chemopreventive agent, Se-methylselenocysteine, has focused on recent studies in generating new and more efficacious selenium chemopreventive compounds for mammary tumorigenesis. Future studies are needed to determine whether this compound is also an improvement over existing selenium compounds for prostate, colorectal, and lung cancer. These studies are particularly important, because the pioneering study of Clark et al. (1) has been translated into a large National Cancer Institute-sponsored intervention study on prostate cancer. If the new intervention study supports the earlier studies and the new animal studies demonstrate that Se-methylselenocysteine is effective in other tumor systems, then future studies can take advantage of Se-methylselenocysteine as a more efficacious agent. The renewed interest in selenium as chemoprevention compounds is related directly to Clark's landmark study. We, as contemporary investigators of selenium-mediated chemoprevention, are deeply indebted to Larry Clark's dedication, perseverance, and foresight.

Acknowledgments and Notes

The work from the authors' laboratories was supported by National Cancer Institute Grant P01-CA-45164. Address correspondence to Daniel Medina, Dept. of Molecular and Cellular Biology, Baylor College of Medicine, Houston, TX 77030. Phone: (713) 798-4483. FAX: (713) 790-0545. E-mail: dmedina@bcm.tmc.edu.

Submitted 3 October 2000; accepted in final form 30 November 2000.

References

1. Clark LC, Combs GF, Turnbull BW, Slate EH, Chalker DK, et al.: Effects of selenium supplementation for cancer prevention in patients with carcinoma of the skin. *JAMA* **276,** 1957–1985, 1996.
2. Shamberger RJ, Tytko SA, and Willis CE: Antioxidants and cancer. VI. Selenium and age-adjusted human cancer mortality. *Arch Environ Health* **31,** 231–235, 1976.
3. Clark LC, Cantor K, and Allaway WH: Selenium in forage crops and cancer mortality in US counties. *Arch Environ Health* **46,** 37–42, 1991.
4. Salonen JT, Salonen R, Lappetelainen R, Maenpaa PH, Alfthan G, et al.: Risk of cancer in relation to serum concentrations of selenium and vitamins A and E: matched case-control analysis of prospective data. *Br Med J* **290,** 417–420, 1985.
5. Clark LC, Hixson LJ, Combs GF Jr, Reid ME, Turnbull BW, et al.: Plasma selenium concentration predicts the prevalence of colorectal adenomatous polyps. *Cancer Epidemiol Biomarkers Prev* **2,** 41–46, 1993.
6. Ip C and Medina D: Current concepts on selenium and mammary tumorigenesis. In *Cell and Molecular Biology of Experimental Breast Cancer*, Medina D, Hepper GH, Kidwell WR, and Anderson E (eds). New York: Plenum, 1987, pp 479–494.
7. Medina D and Morrison DG: Current ideas on selenium as a chemopreventive agent. *Pathol Immunopathol Res* **7,** 187–199, 1988.
8. Medina D and Lane HW: Stage specificity of selenium-mediated inhibition of mouse mammary tumorigenesis. *Biol Trace Elem Res* **5,** 297–306, 1983.
9. Shrauzer GN, White DA, and Schneider CJ: Inhibition of the genesis of spontaneous mammary tumors in C3H mice: effects of selenium and of selenium-antagonistic elements and their possible role in human breast cancer. *Bioinorg Chem* **6,** 265–270, 1976.
10. Ganther HE: Pathways of selenium metabolism including respiratory excretory products. *J Am Coll Toxicol* **5,** 1–5, 1986.
11. Ip C and Ganther HE: Relationship between the chemical form of selenium and anticarcinogenic activity. In *Cancer Chemoprevention*, Wattenberg L, Lipkin M, Boone CW, and Kelloff GJ (eds). Boca Raton, FL: CRC, 1992, pp 479–488.
12. Ip C and Ganther HE: Activity of methylated forms of selenium in cancer prevention. *Cancer Res* **50,** 1206–1211, 1990.
13. Ip C, Thompson HJ, Zhu Z, and Ganther HE: In vitro and in vivo studies of methylseleninic acid: evidence that a monomethylated selenium metabolite is critical for cancer chemoprevention. *Cancer Res* **60,** 2882–2886, 2000.
14. Sinha R, Unni E, Ganther HE, and Medina D: Methylseleninic acid, a potent growth inhibitor of synchronized mouse mammary epithelial tumor cells in vitro. *Biochem Pharmacol* **61,** 311–317, 2000.
15. Ip C, Zhu Z, Thompson HJ, Lisk D, and Ganther HE: Chemoprevention of mammary cancer with Se-allylselenocysteine and other selenoamino acids in the rat. *Anticancer Res* **19,** 2875–2880, 1999.
16. Andreadou I, Menge WMPB, Commandeur JNM, Worthington EA, and Vermeulen NPE: Synthesis of novel Se-substituted selenocysteine derivatives as potential kidney selective prodrugs of biologically active selenol compounds: evaluation of kinetics of β-elimination reactions in rat renal cytosol. *J Med Chem* **39,** 2040–2046, 1996.
17. Ip C, Lisk DJ, and Thompson HJ: Selenium-enriched garlic inhibits the early stage but not the late stage of mammary carcinogenesis. *Carcinogenesis* **17,** 1979–1982, 1996.
18. Allan CB, Lacourciere GM, and Stadtman TC: Responsiveness of selenoproteins to dietary selenium. *Annu Rev Nutr* **19,** 1–16, 1999.
19. Mustacich D and Powis G: Review article: thioredoxin reductase. *Biochem J* **346,** 1–8, 2000.
20. Foiles PG, Fujiki H, Suganuma M, Okabe S, Yatsunami J, et al.: Inhibition of PKC and PKA by chemopreventive organoselenium compounds. *Int J Oncol* **7,** 685–690, 1995.
21. Sinha R, Kiley SC, Lu J, Thompson HJ, Moraes R, et al.: Effect of methyl-selenocysteine on protein kinase C activity in inhibition of synchronous mouse mammary tumor cells in vitro. *Cancer Lett* **146,** 135–145, 1999.
22. Gopalakrishna R, Gundimeda U, and Chen Z-H: Cancer-preventive selenocompounds induce a specific redox modification of cysteine-rich regions in Ca^{2+}-dependent isoenzymes of protein kinase C. *Arch Biochem Biophys* **348,** 25–36, 1997.
23. Gopalakrishna R, Chen Z-H, and Gundimeda U: Selenocompounds induce a redox modulation of protein kinase C in the cell, compartmentally independent from cytosolic glutathione: its role in inhibition of tumor promotion. *Arch Biochem Biophys* **348,** 37–48, 1997.
24. Stapleton SR, Garlock GL, Foellmi-Adams L, and Kletzien RF: Selenium: potent stimulator of tyrosyl phosphorylation and activator of MAP kinase. *Biochim Biophys Acta* **1355,** 259–269, 1997.
25. Sohn OS, Fiala ES, Upadhyaya P, Chae Y-H, and El-Bayoumy K: Comparative effects of phenylenebis(methylene)selenocyanate isomers on xenobiotic metabolizing enzymes in organs of female CD rats. *Carcinogenesis* **20,** 615–621, 1999.
26. Tanaka T, Makita H, Kawabata K, Mori H, and El-Bayoumy K: 1,4-Phenylenebis(methylene)selenocyanate exerts exceptional chemopreventive activity in rat tongue carcinogenesis. *Cancer Res* **57,** 3644–3648, 1997.
27. Makropoulos V, Brüning T, and Schulze-Osthoff K: Selenium-mediated inhibition of transcription factor NF-κB and HIV-1 LTR promoter activity. *Arch Toxicol* **70,** 277–283, 1996.
28. Adler V, Pincus MR, Posner S, Upadhyaya P, El-Bayoumy K, et al.: Effects of chemopreventive selenium compounds on Jun N-kinase activities. *Carcinogenesis* **17,** 1849–1854, 1996.

29. Kaeck M, Junxuan L, Strange R, Ip C, Ganther HE, et al.: Differential induction of growth arrest inducible genes by selenium compounds. *Biochem Pharmacol* **53**, 921–926, 1997.

30. Lanfear J, Fleming J, Wu L, Webster G, and Harrison PR: The selenium metabolite selenodiglutathione induces p53 and apoptosis: relevance to the chemopreventive effects of selenium? *Carcinogenesis* **15**, 1387–1392, 1994.

31. Spyrou G, Bjornstedt M, Kumar S, and Holmgren A: AP-1 DNA-binding activity is inhibited by selenite and selenodiglutathione. *FEBS Lett* **368**, 59–63, 1995.

32. Fiala ES, Staretz ME, Pandya GA, El-Bayoumy K, and Hamilton SR: Inhibition of DNA cytosine methyltransferase by chemopreventive selenium compounds, determined by an improved assay for DNA cytosine methyltransferase and DNA cytosine methylation. *Carcinogenesis* **19**, 597–604, 1998.

33. Cao YZ, Reddy CC, and Sordillo LM: Altered eicosanoid biosynthesis in selenium-deficient endothelial cells. *Free Radic Biol Med* **28**, 381–389, 2000.

34. Lu J, Kaeck M, Jiang C, Wilson AC, and Thompson HJ: Selenite induction of DNA strand breaks and apoptosis in mouse leukemic L1210 cells. *Biochem Pharmacol* **47**, 1531–1535, 1994.

35. Lu J, Jiang C, Kaeck M, Ganther H, Vadhanavikit S, et al.: Dissociation of the genotoxic and growth inhibitory effects of selenium. *Biochem Pharmacol* **50**, 213–219, 1995.

36. Lu J, Pei H, Ip C, Lisk DJ, Ganther H, et al.: Effect on an aqueous extract of selenium-enriched garlic on in vitro markers and in vivo efficacy in cancer prevention. *Carcinogenesis* **17**, 1903–1907, 1996.

37. Thompson HJ, Wilson A, Lu J, Singh M, Jiang C, et al.: Comparison of the effects of an organic and an inorganic form of selenium on a mammary carcinoma cell line. *Carcinogenesis* **15**, 183–186, 1994.

38. Wilson AC, Thompson HJ, Schedin PJ, Gibson NW, and Ganther HE: Effect of methylated forms of selenium on cell viability and the induction of DNA strand breakage. *Biochem Pharmacol* **43**, 1137–1141, 1992.

39. Sinha R and Medina D: Inhibition of cdk2 kinase activity by Se-methylselenocysteine in synchronized mouse mammary epithelial cells. *Carcinogenesis* **18**, 1541–1547, 1997.

40. Ip C, Thompson HJ, and Ganther HE: Selenium modulation of cell proliferation and cell cycle biomarkers in normal and premalignant cells of the rat mammary gland. *Cancer Epidemiol Biomarkers Prev* **9**, 49–54, 2000.

41. Jiang C, Jiang W, Ip C, Ganther H, and Lu J: Selenium-induced inhibition of angiogenesis in mammary cancer at chemopreventive levels of intake. *Mol Carcinog* **26**, 213–225, 1999.

42. Ganther HE: Review: selenium metabolism, selenoproteins and mechanisms of cancer prevention: complexities with thioredoxin reductase. *Carcinogenesis* **20**, 1657–1666, 1999.

43. Bansal M, Ip C, and Medina D: Levels of ^{75}Se labeling of specific proteins as a consequence of dietary selenium concentration in mice and rats. *Proc Soc Exp Biol Med* **196**, 147–154, 1991.

44. Behne D, Hilmert H, Schied S, Gessner H, and Elger W: Evidence for specific selenium target tissues and new biologically important selenoproteins. *Biochim Biophys Acta* **966**, 12–21, 1988.

NUTRITION AND CANCER, *40*(1), 18–27

Multiorgan Sensitivity to Anticarcinogenesis by the Organoselenium 1,4-Phenylenebis(Methylene)Selenocyanate

Karam El-Bayoumy, C. V. Rao, and Bandaru S. Reddy

Abstract: The data in this report clearly indicate that the form (structure) in which selenium is used is the most critical determinant of success in future clinical trials. Synthetic organoselenium compounds can be tailored to achieve greater chemopreventive efficacy with minimal toxic side effects by structural modifications. We demonstrated that 1,4-phenylenebis(methylene)selenocyanate is a powerful chemopreventive agent against the development of experimental colon, mammary, lung, and oral carcinogenesis. On the basis of metabolism studies of organoselenium compounds and those reported in the literature, our working hypothesis is that aromatic selenol intermediates are important entities in cancer chemoprevention. In addition, we suggest that 1,4-phenylenebis(methylene)selenocyanate not only serves as a chemopreventive agent, but it may be valuable in preventing metastatic diseases in future studies in the clinic.

Introduction

To elucidate fully how dietary modifications can be effectively harnessed for cancer control, a stepwise approach must be taken. Toward this end, diet modifications and chemoprevention constitute valuable and plausible approaches (1–3). Thus we are searching for optimal diets and for naturally occurring agents in routinely consumed foods that may inhibit cancer development. Structural modifications of established, naturally occurring chemopreventive agents have led to synthetic agents with even greater efficacy and lower toxicity. The broadly aimed chemoprevention research program in our laboratories at the American Health Foundation uses in vitro and in vivo preclinical assays (4,5). Predictable incidences of tumors are induced by agents such as those present in tobacco products or food or, in some instances, by synthetic carcinogens to provide a yardstick for measuring chemopreventive efficacy of naturally occurring or newly developed synthetic compounds.

Knowledge gained from epidemiological studies, although sometimes ambiguous, suggests that an increased risk for certain human diseases, including cancer, is related to insufficient intake of selenium (6–8). The use of selenium in human clinical trials is limited. The late Dr. Clark, who directed the Nutritional Prevention of Cancer Projects at the Arizona Cancer Center, had focused on the beneficial health effects of increased selenium intake in humans and had conducted epidemiological studies in this area (8). A key achievement is the double-blind, randomized trial of selenium-enriched yeast in patients with nonmelanoma skin cancer that led to the unexpected discovery that selenium protects against colon, lung, and prostate cancers (9,10). The outcome of Clark's trial stimulated the initiation of two new clinical intervention trials in three European countries (PRECISE) and in the United States (SELECT) (11).

Preclinical investigations had led to a systematic study of selenium compounds as one group of cancer-chemopreventive agents that merits further research (12). At doses well above the physiological requirement, inorganic selenium in the diet or in the drinking water protects laboratory animals against cancer of the mammary gland, colon, lung, pancreas, liver, and skin (4). Inorganic and some naturally occurring selenium containing amino acids, such as selenomethionine and selenocysteine, were equally effective chemopreventive agents and had comparable toxicity (12). Yet this toxicity inhibited further research until we introduced novel synthetic organoselenium compounds (4). The rationale for synthesizing these agents is described in detail elsewhere (4,5).

Synthetic organoselenium compounds were examined as chemopreventive agents in several animal tumor models and, in some instances, were compared with inorganic selenite (Fig. 1). Achieving optimal chemopreventive potency with lowest toxicity continues to be a primary goal in our program to develop organoselenium chemopreventive agents. This report is not meant to provide a comprehensive review of the subject but, rather, to provide basic information on the efficacy of specific organoselenium compounds in several animal tumorigenesis model systems and their potential role in preventing metastatic diseases. The metabolism and mechanisms that may underlie cancer chemoprevention, as well as future directions, are discussed.

The authors are affiliated with the American Health Foundation, Valhalla, NY 10595.

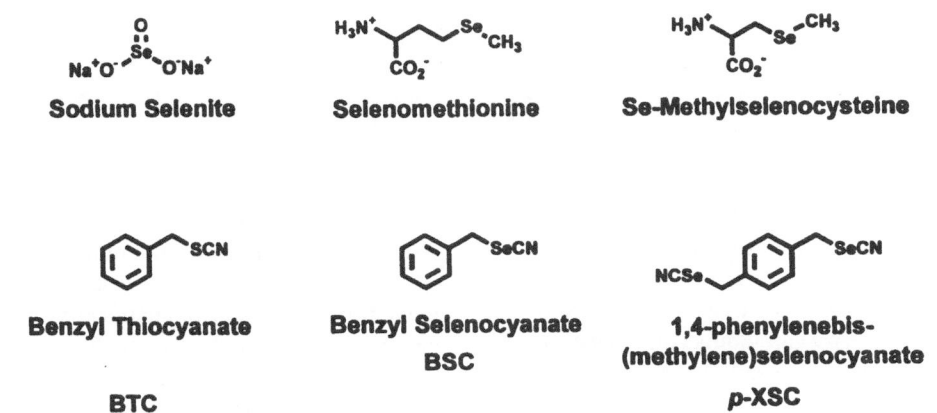

Figure 1. Structures of selenium and sulfur chemopreventive agents.

Chemoprevention of Mammary Cancer by Organoselenium Compounds

Breast cancer, one of the major health problems, is the second most frequent cause of cancer-related deaths in women. Each year, breast cancer is diagnosed in 910,000 women worldwide and 376,000 women die from the disease (reviewed in Ref. 13). Increasingly, available molecular and biochemical tools, as adjuncts to epidemiology, have provided a better understanding of the process of cancer formation or carcinogenesis. Such understanding undoubtedly will enable us to define the cancer risk of individuals and to modulate this risk by means of agents that can alter critical steps in the carcinogenesis process. The etiology of most breast cancer remains obscure, and there are no established primary prevention strategies. Moreover, advances in therapy are limited, and alternatives need to be developed for breast cancer control. Thus began the search for synthetic or naturally occurring agents that can inhibit the neoplastic events that precede the occurrence of clinically detectable cancers. We describe here highlights of our efforts on the role of organoselenium compounds as chemopreventive agents in the breast tissue in preclinical studies.

Benzyl selenocyanate (BSC) represents the first generation of synthetic organoselenium compounds developed in our laboratory (Fig. 1). BSC inhibited dimethylbenz[a]anthracene (DMBA)-induced mammary tumors in rats; its sulfur analog benzyl thiocyanate (BTC) and selenite had no effect during the initiation phase of carcinogenesis (5). The presence of selenium in BSC is necessary for its chemopreventive activity, as is evident from studies comparing BSC with its sulfur analog BTC. Further support for the requirement of selenium in cancer chemoprevention was obtained from a recent study showing that diallyl selenide is superior to diallyl sulfide in inhibiting mammary cancer induced by DMBA (14).

Although BSC was more effective and less toxic than sodium selenite, subchronic toxicity studies in rats indicated that the body weights of rats given dietary BSC or selenite were significantly lower than those in control animals, suggesting that these agents exerted a systemic toxic effect or induced avoidance of food (5). Therefore, we embarked on an effort to design more effective organoselenium compounds with minimal toxic side effects. Various structural analogs of BSC were synthesized. We found that 1,4-phenylenebis(methylene)selenocyanate (p-XSC), representing the second generation of synthetic organoselenium compounds, was markedly less toxic than BSC (5). We then assessed the efficacy of dietary p-XSC (40 ppm as selenium) against DMBA-induced mammary carcinogenesis during the initiation phase (15). Finding that p-XSC inhibited DMBA-induced mammary carcinogenesis prompted us to examine the effect of dietary p-XSC on DMBA-DNA binding in the liver and mammary glands under conditions identical to those in the bioassay. We found that dietary p-XSC inhibited total DMBA-DNA binding in the mammary glands but not in the liver (15). The most profound effect was observed 24–48 h after DMBA administration. The inhibition of binding was attributed to a reduction in the formation of the three major adducts derived from bay-region diol epoxides of DMBA (anti-diol epoxide:deoxyguanosine, syn-diol epoxide:deoxyadenosine, and anti-diol epoxide:deoxyadenosine). Although details of the mechanism are not clear, inhibition of DMBA-DNA binding in the target tissue provides a plausible explanation for the chemopreventive effects of p-XSC during the initiation phase of carcinogenesis. In a recent report, we demonstrated that p-XSC had no effect on DMBA-induced oxidative damage measured as 8-hydroxy-2'-deoxyguanosine (16). This result suggests that DMBA-induced DNA covalent modification is necessary for the initiation of mammary carcinogenesis.

In a follow-up study, we extended these earlier investigations by testing the response to lower levels of p-XSC (5, 10, and 15 ppm as selenium) when given before or after DMBA administration (17). The results clearly indicated that the chemopreventive activity was not limited to the initiation stage of carcinogenesis but was also exerted at postinitiation. In this study, the tumor inhibitor activities of methyl selenocyanate, BSC, and p-XSC were also compared with those of potassium selenocyanate and selenite. Cancer inhibition was ranked by calculating the ratio of the maximum tolerable dose to the effective dose that produced ~50% inhibition of total tumor yield; this ratio is referred to as the chemopreventive index. The chemopreventive index of p-XSC was estimated as 4, while the indexes of BSC, methyl selenocyanate, potassium selenocyanate, and selenite were 2.5, 2.0, 1.3, and 1.3, respectively (17); the chemopreventive index of selenite is comparable to that of selenomethionine in this animal model (4,5).

To study the possible mechanism(s) for inhibition of postinitiation events of carcinogenesis by p-XSC, Thompson et al. (18) compared the effects of p-XSC and sodium selenite, an established chemopreventive agent shown to be a growth inhibitor in mammary carcinoma cell lines of mice (19). Treatment with p-XSC caused a three- to sixfold greater accumulation of selenium within cells than did treatment with equivalent amounts of selenite; also, cells were able to tolerate higher levels of selenium when it was derived from p-XSC (18). Both compounds effected a dose-dependent reduction in cell number after 24 h of exposure and a dose-dependent increase in cell death by apoptosis. Results obtained by us and by Thompson et al. confirmed that the effect of p-XSC on apoptosis was more pronounced than that of selenite (18,20). Inhibition of cell growth and induction of apoptosis may partially account for the chemopreventive properties of selenite and p-XSC. We also compared the effects of p-XSC and selenite on mammary tumor cell lines derived from humans and rats (20,21); p-XSC was indeed capable of inhibiting thymidine kinase, while equal concentrations of selenium in the form of selenite had no effect. Adler et al. (22) demonstrated that p-XSC, but not selenite, can modulate Jun NH_2-terminal kinase activities in such a manner as to contribute to the cell's ability to cope with stress such as ultraviolet light exposure. Collectively, we can state that p-XSC is more effective than selenite, selenomethionine, and other forms of synthetic organoselenium compounds developed in our laboratory as a cancer-chemopreventive agent against mammary cancer in model assays with rodents. Our current efforts are focused on the identification of the precise form (metabolite) of selenium that is responsible for the chemopreventive effects of organoselenium compounds, especially p-XSC.

Chemoprevention of Colon Cancer by Organoselenium Compounds

Large bowel cancer is one of the most common and persistent human malignancies in the western world, including the United States. Globally, >875,000 men and women were afflicted with this cancer, and >510,000 individuals died of colon cancer in 1996 (23). The American Cancer Society estimates that, in the United States, there will be 137,000 new cases and ~57,000 deaths attributable to this cancer alone in the year 2000; thus large bowel cancer is a major public health problem (24). Chemoprevention has the potential to be a major component of colorectal cancer control. Rapidly evolving progress in chemoprevention research has brought about innovative approaches to the prevention of colorectal cancer. Prevention strategies that embrace intervention by

nutritional modification and chemopreventive agents that can retard, block, or reverse the process of colorectal carcinogenesis or reduction of underlying factors can be applied across a continuum of the general population. In this connection, it is noteworthy that Clark et al. (9) reported results from a randomized clinical trial in which they found that supplementation of selenium-enriched brewer's yeast reduced the incidence and mortality from cancer of the colon. This is corroborated by studies with selenium supplementation of the diet in chemically induced colon carcinogenesis in preclinical models (5).

Substantial efforts were made in our institute to find and/or develop forms of organic selenium compounds that have maximal chemopreventive efficacy against colon cancer and low possible toxicity. During the past decade, preclinical efficacy studies conducted in our laboratory have indicated that certain synthetic organoselenium compounds hold great promise as chemopreventive agents against colon cancer, because they have been found to be superior to historically used selenium compounds, such as sodium selenite and selenomethionine (25–28). Among the organoselenium compounds developed, p-XSC was found to be less toxic than sodium selenite and selenomethionine (4,5,17).

Experiments conducted in our laboratory have shown that administration of p-XSC in a high-fat diet at dose levels of 20 and 40 ppm equivalent to 10 and 20 ppm as selenium significantly suppressed azoxymethane (AOM)-induced colon carcinogenesis in male Fischer 344 (F344) rats at initiation and postinitiation stages (26). Inhibition of colon carcinogenesis in this preclinical model was associated with inhibition of prostaglandin E_2 levels in the colonic mucosa and tumors (26). Also, p-XSC enhanced glutathione peroxidase activity in the colon (26). Because of the known colon tumor-promoting effects of high dietary fat, we also examined whether fat intake would have an impact on the chemopreventive efficacy of p-XSC against AOM-induced colon carcinogenesis in F344 rats (27). This protocol is very important, because secondary prevention of colon cancer by administration of chemopreventive agents alone may reduce the risk of colon cancer somewhat among high-risk individuals consuming a Western-style diet, but it is likely to be most effective along with lifestyle changes, such as maintaining a low intake of dietary fat. Results of this study indicate that administration of a high-fat diet and p-XSC during or after the initiation period significantly decreased the incidence of adenocarcinomas (27). The incidence and multiplicity of noninvasive adenocarcinomas were significantly inhibited when p-XSC was administered during the postinitiation period (Table 1). Also, significant suppression of the incidence and multiplicity of adenocarcinomas, specifically noninvasive carcinomas, was observed in rats fed the p-XSC-containing low-fat diet during the postinitiation period compared with the findings for rats fed the low-fat diet alone. It is interesting that administration of p-XSC, along with a low-fat diet, during the postinitiation period reduced the incidence and multiplicity of adenocarcinomas by 70%

Table 1. Chemopreventive Efficacy of *p*-XSC Administered in Low- or High-Fat Diets During Postinitiation Stage of AOM-Induced Colon Carcinogenesis in Male F344 Rats[a,b]

Experimental Groups	Colon Tumor Incidence, % animals with tumors			Colon Tumor Multiplicity, no. of tumors/animals		
	Noninvasive	Invasive	Total[c]	Noninvasive	Invasive	Total
High-fat diet						
Control	69	31	90	1.24 ± 1.15	0.38 ± 0.62	1.62 ± 1.05
20 ppm *p*-XSC	37*	23	52*	0.53 ± 0.62	0.27 ± 0.52	0.80 ± 0.85
Low-fat diet						
Control	50	10	53*	1.77 ± 0.97	0.10 ± 0.31	0.87 ± 1.07*
20 ppm *p*-XSC	10[†,‡]	17	27[†,‡]	1.10 ± 0.31	0.20 ± 0.48	0.30 ± 0.53[†,‡]

a: Colon tumor multiplicity values are means ± SE; $n = 29$–30. *p*-XSC, 1,4-phenylenebis(methylene)selenocyanate; AOM, azoxymethane; F344, Fischer 344.

b: Statistical significance is as follows: *, Significantly different from high-fat control diet group ($P < 0.01$–0.02); †, significantly different from high-fat control diet group ($P < 0.00002$–0.000002); ‡, significantly different from low-fat control diet group ($P < 0.03$–0.004).

c: Total includes noninvasive + invasive adenocarcinomas.

and 81.5%, respectively, compared with the results obtained for the high-fat control group. This inhibition was again strongest in terms of noninvasive adenocarcinomas. The fact that the administration of *p*-XSC with low- or high-fat dietary regimens during the postinitiation phase significantly suppressed the incidence and multiplicity of noninvasive adenocarcinomas suggests that this compound can effectively inhibit the progression from noninvasive to invasive tumors. This finding has practical implications in the chemoprevention of colon cancer. These studies provide convincing evidence that the chemopreventive efficacy is enhanced when it is administered in a low-fat diet. The chemopreventive indexes for *p*-XSC when administered in high- and low-fat diets were 2.25 and 3.35, respectively. The high index values for *p*-XSC signify that this compound is well tolerated at doses required for chemoprevention. Also, the high index value for *p*-XSC when administered in a low-fat diet (3.35) compared with its high-fat counterpart (2.25) provides evidence that the chemopreventive effect of this agent is enhanced when it is given in a low-fat diet. We also observed that administration of *p*-XSC in low- and high-fat diets increases apoptosis in colon tumors at a rate that parallels the colon tumor inhibition by this agent (29) (Table 1). Also the maximal induction of apoptosis was achieved in colon tumors of animals fed *p*-XSC along with a low-fat diet compared with the tumors of animals fed *p*-XSC along with a high-fat diet. This finding underscores the concept that a reduction in dietary fat may offer an important adjunct to chemopreventive efficacy in human colorectal cancer prevention trials involving organoselenium compounds. This concept should be explored in a broad range of human trials involving other chemopreventive agents.

The progression from normal epithelium to colon cancer is a multistep process involving accumulation of multiple genetic alterations. Adenomatous polyposis coli (*APC*), a tumor suppressor gene on chromosome 5q, is considered to be a gatekeeper in colon tumorigenesis. Its inactivation leads to the development of adenomatous polyps (30). In most cases of colonic neoplasia, both inherited and somatic, the *APC* gene is mutationally inactivated by the introduction of premature stop codons and/or deleted by loss of heterozygosity

(31). Several mouse lineages that are heterozygous for a specific mutation at the endogenous *APC* gene have been developed and characterized with respect to their intestinal multiple tumors. Among mutant *APC* mice, *APC^min*, an inbred strain of C57BL/6J mice, carries a germline truncation of one *APC* allele (30); these mice develop multiple intestinal adenomas by 4 mo of age. This model has been utilized to unravel the basic mechanisms of intestinal tumor formation and to test cancer-chemopreventive agents. Studies of the biochemical mechanisms downstream of the *APC* mutation have also provided important leads. Overexpression of cyclooxygenase (COX)-2 is one of the most significant observations in this respect. Oshima et al. (32) elegantly provided definitive evidence that induction of COX-2 is an early, rate-limiting step for adenoma formation. Recent studies also suggest another mechanism for *APC* function, namely, that *APC* binds to β-catenin, which in turn binds to and regulates the transcription factor lymphoid enhancer-binding factor-1 (30,33). We have evaluated the chemopreventive efficacy of *p*-XSC at dose levels of 5 and 10 ppm as selenium in *APC^min* mice (34). Administration of *p*-XSC in the diet significantly decreased the rate of formation of small intestinal tumors ($P < 0.0001$) and colon tumors in *APC^min* mice (Fig. 2). *p*-XSC produced a dose-dependent inhibition of tumors in small intestine and colon. Small intestinal and colonic mucosa and tumors were also assayed for β-catenin, COX-2 expression, and COX isoform activities. Mice fed 20 ppm *p*-XSC had significantly lower levels of β-catenin expression and COX-2 activity in polyps (Table 2). These observations demonstrate that the synthetic organoselenium compound *p*-XSC possesses antitumor activity against genetically predisposed neoplastic lesions, such as familial adenomatous polyposis. Although the exact mechanism(s) for this antitumor activity of *p*-XSC remains to be elucidated, it appears that modulation of β-catenin expression and COX-2 activity is associated with inhibition of intestinal polyps.

In summary, an impressive body of observation supports the concept that synthetic organoselenium compounds are key modulators of colon cancer. Accumulating evidence indicates that the synthetic organoselenium compound *p*-XSC

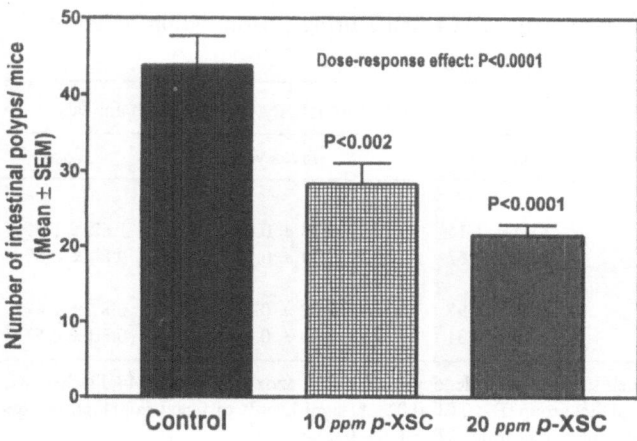

Figure 2. Effect of p-XSC on intestinal tumor formation in *APCmin* mice. Six-week old male (heterozygous) C57BL/6J-*APCmin* mice were fed high fat diets containing 0, 10, or 20 ppm p-XSC. After 80 days, the mice were sacrificed and their intestines were excised and evaluated for polyps.

Table 2. Effect of p-XSC on COX-2 Activity and β-Catenin Expression in Intestinal Tumors of *APCmin* Mice[a,b]

Experimental Groups	COX-2 Activity	β-Catenin Expression
Control	58.1 ± 4.0	274 ± 24
20 ppm p-XSC	41.2 ± 3.8*	128 ± 14[†]

a: Values are means ± SE, expressed in pmol [^{14}C]15R-hydroxy-eicosatetraenoic acid formed/mg protein/20 min [for cyclooxygenase (COX)-2 activity] and ng β-catenin/mg protein by Western blot analysis. *APC*, adenomatous polyposis coli gene.

b: Statistical significance is as follows: *, significantly different from control diet group, $P < 0.03$; †, significantly different from control diet group, $P < 0.0001$.

holds great promise as a chemopreventive agent, because it is far less toxic, yet more effective, than inorganic selenium and selenomethionine in inhibiting colon carcinogenesis. More importantly, the chemopreventive efficacy of this agent is more pronounced when given along with a low-fat diet, thus making a strong case for the use of low-fat diet regimens along with a chemopreventive agent as a desirable approach for primary prevention in the general population and for secondary prevention of colon cancer in high-risk individuals.

Chemoprevention of Lung Cancer by Organoselenium

Effective control and prevention of tobacco consumption and the reduction of environmental exposure to known carcinogens remain major goals for cancer prevention (2,35). Although lung cancer mortality rates among white men have not further increased in the United States since 1985 because of primary prevention efforts and product modification, lung cancer mortality throughout the world will remain very high for many years (36,37). To reduce these epidemic propor-

tions of lung cancer, chemoprevention, although still in a phase of extensive development, offers a plausible approach. The results achieved so far in this area of research are promising. However, chemoprevention should never be considered a substitute for primary prevention efforts, but a complementary approach. For example, for persons who have already quit smoking, chemopreventive agents can potentially reduce lung cancer risk, which remains high 5–10 yr after cessation of smoking. Therefore, we have set up model studies on lung cancer induction and prevention in laboratory animals. The carcinogen in this case is 4-(methylnitrosamino)-1-(3-pyridyl)-1-butanone (NNK), a nicotine-derived nitrosamine present in tobacco and tobacco smoke; NNK is strongly implicated in the pathogenesis of tobacco-related lung cancer in humans (38).

The well-established A/J mouse lung adenoma assay was chosen for these investigations, since the lung is the major target organ for NNK-induced tumorigenesis (38). Bioassays testing the efficacy of selenite on the inhibition of tumor induction by nitrosamines and nitrosamides are scarce (4). The results of these limited assays generally indicate that, as a chemopreventive agent, selenite is either ineffective or extremely weak (reviewed in Ref. 4). Prior to our report, selenite had not been tested in the NNK model (39). The chemopreventive effect of 5, 10, and 15 ppm selenium given as p-XSC on lung tumor induction by NNK was examined in female A/J mice by administration of p-XSC in the diet (39). Sodium selenite (5 ppm selenium) was given in the same manner. Mice were fed an experimental diet, containing the selenium compounds, 1 wk before intraperitoneal injection of 10 μmol of NNK in 0.1 ml of saline and throughout the experiment until termination at 16 wk after carcinogen administration. p-XSC significantly lowered lung tumor multiplicity from 7.6 tumors per mouse in the control group to 4.1, 3.3, and 1.8 tumors per mouse in animals given 5, 10, or 15 ppm selenium. In contrast, 5 ppm sodium selenite had no protective effect against lung tumor multiplicity. A further investigation was designed to determine whether p-XSC inhibits pulmonary neoplasia induced by NNK in female A/J mice during the initiation phase of carcinogenesis or during the postinitiation phase (40); the naturally occurring selenomethionine was also included in this study. The results clearly indicate that selenomethionine had no effect, but p-XSC effectively blocked and suppressed NNK-induced lung tumor development in mice (40).

In an effort to further delineate the mechanism of action of p-XSC inhibition of NNK-induced lung neoplasms, we examined its effect on NNK-induced DNA methylation in the lungs and liver of A/J mice and rats (41). The effect of selenite was also investigated in this study. We found that p-XSC (15 ppm selenium) reduced 7-methylguanine formation in this model assay by ~50% after 4 h. Dietary p-XSC (5 or 15 ppm selenium) also inhibited formation of O^6-methylguanine and 7-methylguanine in the lung; however, selenite had no effect. Because p-XSC suppressed NNK-DNA methylation in the rat lung (41), one may assume that it would also protect against NNK-induced lung carcinogenesis in rats; this will be explored in the near future. The preliminary findings suggest that the inhibition of DNA adduct formation in the lung may, in part, account for the chemopreventive effect of p-XSC. p-XSC inhibited NNK-induced 8-hydroxy-2′-deoxyguanosine in rat and mouse lung (42). Whether reduction of lung cancer in the study conducted by Clark et al. (9), which employed selenium-enriched yeast, is due to reduction in levels of genetic damage needs to be determined. In a recent report, we demonstrated that p-XSC inhibited lung tumors induced by a mixture of NNK and benzo[a]pyrene in A/J mice (43); both carcinogens are considered to be highly relevant to the development of lung cancer in smokers. Our results indicate that p-XSC is a valuable chemopreventive agent against the development of lung tumors induced by these environmental carcinogens to which humans are exposed simultaneously via tobacco smoking.

Chemoprevention of Oral Cancer by Organoselenium

Tobacco and alcohol use are the major risk factors in the development of oral cancer; also, combining exposure to tobacco and alcohol results in increased cancer incidence (44). Overall survival of patients with head and neck cancer remains poor, despite improvements in diagnosis and treatment of oral cancer. Furthermore, a significant number of patients cured of primary tumors will develop a second cancer within a few years, usually in the head and neck region, lung, or esophagus. Many people consuming tobacco and alcohol increase their risk of developing simultaneous or subsequent second primary cancers of these regions (45).

The induction of oral cancer in laboratory animals by the carcinogens 4-nitroquinoline-N-oxide (4-NQO) and DMBA has served as an in vivo model to identify agents that suppress oral carcinogenesis (46). Our study clearly demonstrates that dietary administration of p-XSC to rats during the initiation or postinitiation phase effectively blocks or suppresses 4-NQO-induced oral carcinogenesis without any toxicity and pathological alteration of other organs (47). It is remarkable that no tongue carcinomas developed in rats given the higher dose (15 ppm selenium) of p-XSC after 4-NQO exposure. Feeding of p-XSC also suppressed the development of preneoplastic lesions in oral carcinogenesis. In vivo and in vitro studies suggest that inhibition of cell proliferation of preneoplastic lesions is a fundamental mechanism by which selenium inhibits or delays tumorigenesis (12, 19,21). Feeding of p-XSC during the initiation or postinitiation phase significantly inhibited the expression of cell proliferation biomarkers (bromodeoxyuridine labeling index, number of agyrophilic nucleolar organizer regions, polyamine levels, and ornithine decarboxylase activity). Therefore, one of the mechanisms in the chemopreventive activities of p-XSC may be related to suppression of cell proliferation in the target tissue, especially when the compounds are given during the postinitiation phase. Also, p-XSC feeding affected the activities of glutathione S-transferase and quinone reductase in the liver and tongue (47).

Here, feeding of p-XSC-containing diets (15 ppm selenium) did not cause retardation of body weight gain. No significant pathological alterations in liver, including centrilobular hypertrophy with mild fatty change, were found in rats fed the diet containing 15 ppm selenium formulated as p-XSC during the study. In addition, no pathological abnormalities were seen in other organs, including kidney, lung, heart, testis, prostate, brain, and others. These results confirm the low toxicity of p-XSC. This is important, because the ultimate application of cancer-chemopreventive compounds is that in clinical trials for humans for which long-term parenteral treatment is much less practical than oral administration.

Suppression of Metastasis by Organoselenium Compounds

Metastasis is the most devastating aspect of malignant neoplasms. Although advances in surgery, chemotherapy, and radiotherapy have significantly improved the treatment of primary malignancies, the occurrence of metastasis still leads to poor prognosis and death in patients with malignancy. A complex series of steps is required to permit the successful establishment of tumor metastasis (48,49). Several attempts to find antimetastasis agents, including selenite and selenomethionine, have been made using animal experimental metastasis models (50,51). We examined the modifying effects of dietary p-XSC on experimental metastasis of melanoma cells in a model assay in mice that had been injected with viable B16BL6 melanoma cells, which are syngeneic to C57BL/6 mice (52). The effects of p-XSC on apoptosis in metastatic melanoma cells in the lungs were also examined.

The results indicate that, in mice, diet supplementation with p-XSC reduces pulmonary metastasis of B16BL6 melanoma cells and inhibits the growth of these metastatic tumors in the lung, in part by inducing apoptosis (52). We suggest that p-XSC may be valuable in preventing metastatic diseases in future studies in the clinic. Thus induction of apoptosis may play a critical role in suppression of lung metastasis of melanoma cells by p-XSC. Additional studies are required to delineate the effects of p-XSC on neoangiogenesis and on the expression of matrix metalloproteinases, integrins, nitric oxidase, and metastasis-suppressing gene product. These studies are under way in our laboratories. The elucidation of the mechanisms through which p-XSC exerts its antimetastatic effects represents a fascinating aspect of research in the oncological field.

Role of Metabolism of Organoselenium Compounds in Chemoprevention

Our results and those described in the literature indicate that the chemopreventive efficacy of selenium as an anticarcinogen depends on the chemical form in which it is administered, indicating that metabolism is a prerequisite for cancer prevention (reviewed in Ref. 53). For instance, it is now well established that the antitumor properties of the inorganic form (selenite) are strongly influenced by its metabolism. After absorption, selenite is reduced by thiols (glutathione) and NADPH-dependent reductase through selenodiglutathione (GSSeSG) to the highly toxic H_2Se, which can be incorporated as selenocysteine into selenoproteins such as glutathione peroxidase, type 1 iodothyronine deiodinase, the 57-kDa plasma protein, and selenoprotein P. H_2Se is methylated to mono-, di-, and trimethylated derivatives before excretion. Experiments with compounds that are directly converted to the monomethylated form, methyl selenol (54), indicate that the anticarcinogenic effect is directly mediated by methyl selenol or related as yet unidentified intermediates.

As described above, studies in our laboratories have focused on the design of synthetic organoselenium compounds with improved chemopreventive efficacy and lower toxicity than inorganic selenium compounds and certain naturally occurring selenoamino acids (e.g., selenomethionine). Ideally, such agents would be used as dietary supplements to inhibit tumor development caused in different organs by various classes of chemical carcinogens, including those present in the human environment. Understanding the metabolism of organoselenium chemopreventive agents is essential to determine whether the parent compound and/or its metabolites are responsible for chemoprevention.

BSC

In an early study, we demonstrated that [^{14}C]BSC metabolism produces organic selenium intermediates, and the presence of an inorganic selenium metabolite ($Se\overline{CN}$) was also evident (54). Further metabolism of $Se\overline{CN}$ should lead to the same common metabolic pool, i.e., formation of H_2Se and its methylated derivatives, observed with inorganic selenium compounds. Unfortunately, in this study (55) we did not examine the presence of CH_3SeCH_3 in the exhaled air or (CH_3)Se^+ in the urine of rats treated with a single dose of BSC. The results indicate that urine is the major route of excretion. Identification of benzoic and hippuric acids in the urine clearly indicates that BSC is metabolized, in part, via bond cleavage between the benzyl moiety and the selenium atom. The facile reaction of BSC with thiol-containing metabolites in vitro suggests the formation of sulfur-selenium-containing metabolites in vivo (55). The identification of a glutathione conjugate derived from selenocysteine in vivo after the administration of selenocystine to mice was reported (56). Therefore, we hypothesize that the glutathione conjugate of BSC is a primary metabolite mediating the chemopreventive activity by liberating the aromatic selenol moiety (C_6H_5-CH_2-SeH). To obtain preliminary data to support this hypothesis, we initiated a bioassay of limited scope to demonstrate the efficacy of C_6H_5-CH_2-SeSG (5 ppm as selenium) to inhibit DMBA-induced (10 mg/rat) mammary tumors in the rat under conditions identical to those described

for BSC (5). The results (data not shown) suggest that C_6H_5-CH_2-SeSG is better than BSC. Additional studies aimed at determining the effect of BSC and its glutathione conjugate on AOM-induced aberrant crypt foci in F344 rats (57) further support the hypothesis that C_6H_5-CH_2-SeSG is more effective than BSC as an antitumor agent. The results also indicate that C_6H_5-CH_2-SeSG is absorbed in vivo. However, it remained to be determined whether the intact glutathione conjugate or one of its metabolites (C_6H_5-CH_2-SeH or C_6H_5-CH_2-SeSe-CH_2-C_6H_5) can be detected in vivo in biological fluids or can be delivered to extrahepatic tissues in rodents treated with BSC.

p-XSC

In a previous study, we examined the excretion profile of p-[^{14}C]XSC after its oral administration to female CD rats (58). Selenium and radioactivity were monitored. On the basis of radioactivity, 24% of the dose was excreted in the urine and 75% in the feces over 7 days. According to selenium measurement, <1% of the dose was detected in exhaled air. Such a low level of H_2Se (measured as CH_3SeCH_3), in this case, indicates that toxicity can be dissociated from chemopreventive efficacy by structural design of appropriate organoselenium compounds. The identification of tetraselenocyclophane (TSC) as an in vivo metabolite of p-XSC led us to postulate the following metabolic pathway: p-XSC → glutathione conjugate (p-XSeSG) → aromatic selenol moiety (p-XSeH) → TSC. In this pathway, the formation of p-XSeH may be a critical step, since the selenol moiety is considered an important entity in cancer chemoprevention by selenium compounds (53). To test our hypothesis, p-XSeSG and related compounds that can be formed from its further metabolism, i.e., cysteine- and N-acetylcysteine conjugates of p-XSC, were synthesized in our laboratory; comparative efficacy studies between p-XSC and p-XSeSG will follow. Collectively, on the basis of our results and those reported in the literature, we propose Fig. 3, which summarizes metabolic pathways of inorganic and organic selenium compounds. The major focus in cancer prevention by organoselenium compounds in our laboratories is on the design of agents that lead to the formation of those metabolites shown below the dashed line in Fig. 3; selected examples are p-XSeSG, TSC, and p-XSe-CH_3; the latter has been shown to be a promising candidate in cancer chemoprevention (60).

Epilogue

We have clearly demonstrated that synthetic organoselenium compounds can be tailored to achieve greater chemopreventive efficacy with minimal side effects by structural modifications. On the basis of our results, we strongly believe that better candidates than those available in our laboratories can be developed. However, further studies are urgently needed to fully assess the toxicological properties, pharmacokinetics, and mechanism of action of those com-

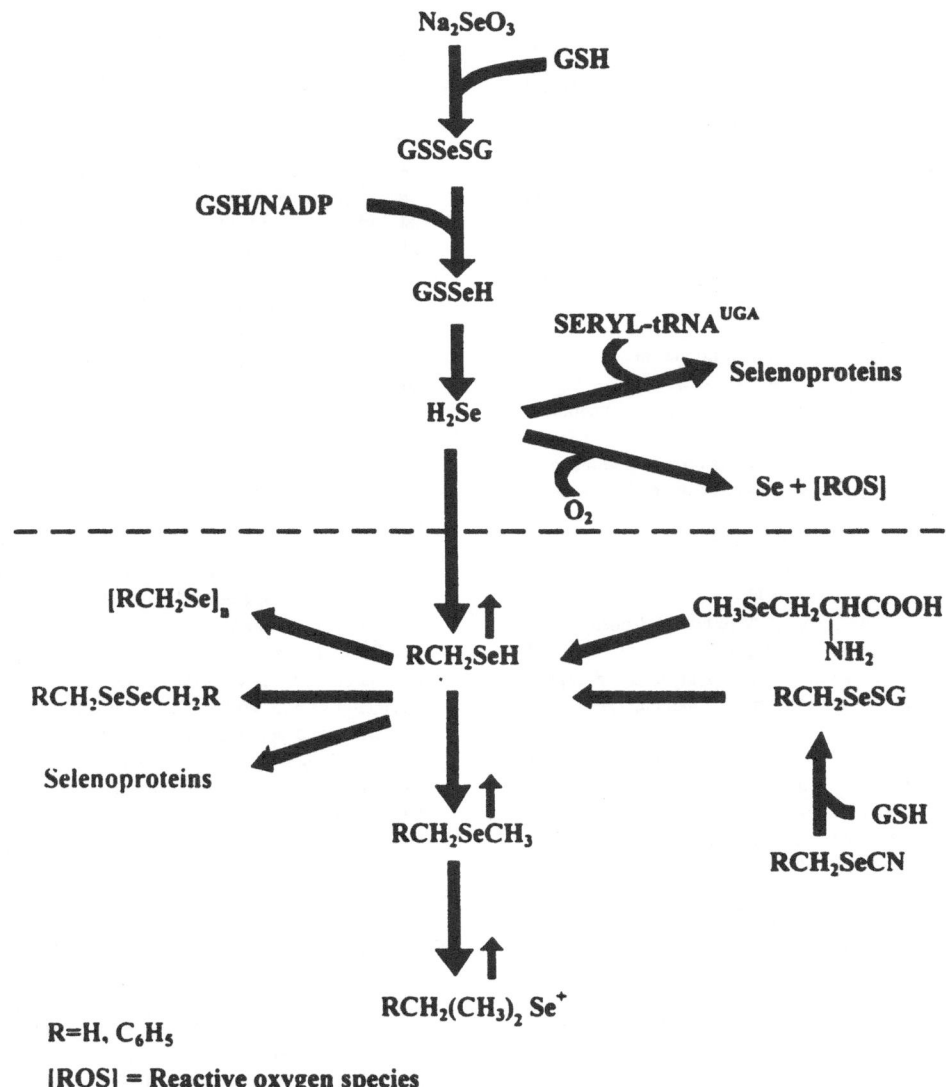

$R=H, C_6H_5$

[ROS] = Reactive oxygen species

Figure 3. Metabolic pathways of inorganic and organic selenium compounds. [Modified from Ref. 59.]

pounds, especially *p*-XSC. The bulk of our knowledge on the mechanisms of action of selenium compounds, including synthetic organoselenium in general, is based, primarily, on animal data and on studies conducted in in vitro systems. How such knowledge is applicable to humans remains unclear because of convincing evidence that some features of selenium metabolism are unique to humans. Clearly, there is a need for pilot studies aimed at determining the role of various forms of selenium on cellular and molecular targets that are critical in the multistep carcinogenic process.

Acknowledgments and Notes

We thank Patricia Sellazzo for preparing and Ilse Hoffmann for editing the manuscript and the staff in the Research Animal Facility for handling and treatment of the animals. The work described in this report was supported by National Institutes of Health Grants CA-46589, CA-70972, and DE-13222. This work is dedicated to the memory of Dr. Larry Clark: a friend, a scientist and, above all, a human being. Address correspondence to Karam El-Bayoumy, Div. of Cancer Etiology and Prevention, American Health Foundation, 1 Dana Rd., Valhalla, NY 10595. E-mail: kelbayoumy@aol.com.

Submitted 15 November 2000; accepted in final form 30 November 2000.

References

1. Doll R: Nature and nurture: possibilities for cancer control. *Carcinogenesis* **17**, 177–184, 1996.
2. El-Bayoumy K, Chung F-L, Richie J Jr, Reddy BS, Cohen L, et al.: Dietary control of cancer. *Proc Soc Exp Biol Med* **216**, 211–223, 1997.
3. Combs GF Jr and Clark LC: Selenium and cancer. In *Nutritional Oncology*, Heber D, Blackburn GL, and Go VLW (eds). San Diego, CA: Academic, 1999, pp 215–222.
4. El-Bayoumy K: The role of selenium in cancer prevention. In *Cancer: Principles and Practice of Oncology*, 4th ed, DeVita VT Jr, Hellman S, and Rosenberg SA (eds). Philadelphia: Lippincott, 1991, vol 2, pp 1–15.
5. El-Bayoumy K, Upadhyaya P, Chae Y-H, Sohn O-S, Rao CV, et al.: Chemoprevention of cancer by organoselenium compounds. *J Cell Biochem Suppl* **22**, 92–100, 1995.
6. Shamberger RJ and Frost DV: Possible protective effect of selenium against human cancer. *Can Med Assoc J* **100**, 682, 1969.

7. Schrauzer GN, White DA, and Schneider CJ: Cancer mortality correlation studies. III. Statistical association with dietary selenium intakes. *Bioinorg Chem* **7**, 35–56, 1977.

8. Clark LC, Cantor KP, and Allaway WH: Selenium in forage crops and cancer mortality in US counties. *Arch Environ Health* **46**, 37–42, 1991.

9. Clark LC, Combs GF Jr, Turnbull BW, Slate EH, Chalker DK, et al.: Effects of selenium supplementation for cancer prevention in patients with carcinoma of the skin: a randomized controlled trial. *JAMA* **276**, 1957–1963, 1996.

10. Clark LC, Dalkin B, Krongrad A, Combs GF Jr, Turnbull BW, et al.: Decreased incidence of prostate cancer with selenium supplementation: results of a double-blind cancer prevention trial. *Br J Urol* **81**, 730–734, 1998.

11. Rayman MP: The importance of selenium to human health. *Lancet* **356**, 233–241, 2000.

12. Ip C: Lessons from basic research in selenium and cancer prevention. *J Nutr* **128**, 1845–1854, 1998.

13. Williams JA and Phillips DH: Mammary expression of xenobiotic metabolizing enzymes and their potential role in breast cancer. *Cancer Res* **60**, 4667–4677, 2000.

14. El-Bayoumy K, Chae Y-H, Upadhyaya P, and Ip C: Chemoprevention of mammary cancer by diallyl selenide, a novel organoselenium compound. *Anticancer Res* **16**, 2911–2916, 1996.

15. El-Bayoumy K, Chae Y-H, Upadhyaya P, Meschter C, Cohen LA, et al.: Inhibition of 7,12-dimethylbenz[a]anthracene-induced tumors and DNA adduct formation in the mammary glands of female Sprague-Dawley rats by the synthetic organoselenium compound, 1,4-phenylenebis(methylene)selenocyanate. *Cancer Res* **52**, 2402–2407, 1992.

16. El-Bayoumy K, Chae Y-H, Rosa JG, Williams LK, Desai D, et al.: The effects of 1-nitropyrene,2-amino-1-methyl-6-phenylimidazo[4,5-b]pyridine and 7,12-dimethylbenz[a]anthracene on 8-hydroxy-2'-deoxyguanosine levels in the rat mammary gland and modulation by dietary 1,4-phenylenebis(methylene)selenocyanate. *Cancer Lett* **151**, 7–13, 2000.

17. Ip C, El-Bayoumy K, Upadhyaya P, Ganther H, Vadhanavikit S, et al.: Comparative effect of inorganic and organic selenocyanate derivatives in mammary cancer chemoprevention. *Carcinogenesis* **15**, 187–192, 1994.

18. Thompson HJ, Wilson A, Lu J, Singh M, Jiang C, et al.: Comparison of the effects of an organic and an inorganic form of selenium on a mammary carcinoma cell line. *Carcinogenesis* **15**, 183–186, 1994.

19. Medina D, Lane HW, and Tracey CM: Selenium and mouse mammary tumorigenesis, an investigation on possible mechanisms. *Cancer Res* **43**, 2460s–2464s, 1983.

20. Tillotson JK, Upadhyaya P, and Ronai Z: Inhibition of thymidine kinase in cultured mammary tumor cells by the chemopreventive organoselenium compound 1,4-phenylenebis(methylene)selenocyanate. *Carcinogenesis* **15**, 607–610, 1994.

21. Ronai Z, Tillotson JK, Traganos F, Darzynkiewicz Z, Conaway CC, et al.: Effects of organic and inorganic selenium compounds on rat mammary tumor cells. *Int J Cancer* **63**, 428–434, 1995.

22. Adler V, Pincus MR, Posner S, Upadhyaya P, El-Bayoumy K, et al.: Effects of chemopreventive selenium compounds on Jun N-kinase activities. *Carcinogenesis* **17**, 1849–1854, 1996.

23. World Health Organization: *The World Health Report 1996*. Geneva, Switzerland: WHO, 1997.

24. Landis SH, Murray T, Bolden S, and Wingo PA: Cancer statistics, 1999. *CA Cancer J Clin* **49**, 9–31, 1999.

25. Nayini JR, Sugie S, El-Bayoumy K, Rao CV, Rigoty J, et al.: Effect of dietary benzylselenocyanate on azoxymethane-induced colon carcinogenesis in male F344 rats. *Nutr Cancer* **15**, 129–139, 1991.

26. Reddy BS, Rivenson A, Kulkarni N, Upadhyaya P, and El-Bayoumy K: Chemoprevention of colon carcinogenesis by the synthetic organoselenium compound 1,4-phenylenebis(methylene)selenocyanate. *Cancer Res* **52**, 5635–5640, 1997.

27. Reddy BS, Rivenson A, El-Bayoumy K, Upadhyaya P, Pittman B, et al.: Chemoprevention of colon cancer by synthetic organoselenium

compounds, 1,4-phenylenebis(methylene)selenocyanate and p-methoxy benzyl selenocyanate, in low and/or high fat fed F344 rats. *JNCI* **89**, 506–516, 1997.

28. Reddy BS, Hirose Y, Lubet RA, Steele VE, Kelloff GJ, et al.: Lack of chemopreventive efficacy of DL-selenomethionine in colon carcinogenesis. *J Mol Med* **5**, 327–330, 2000.

29. Samaha HS, Hamid R, El-Bayoumy K, Rao CV, and Reddy BS: The role of apoptosis in the modulation of colon carcinogenesis by dietary fat and by the organoselenium compound 1,4-phenylenebis(methylene)selenocyanate. *Cancer Epidemiol Biomarkers Prev* **6**, 699–704, 1997.

30. Kinzler KW and Vogelstein B: Cancer-susceptibility genes. Gatekeepers and caretakers. *Nature* **386**, 761–763, 1997.

31. Moser AR, Pitot HC, and Dove WF: A dominant mutation that predisposes to multiple intestinal neoplasia in the mouse. *Science* **247**, 322–324, 1989.

32. Oshima M, Dinchuk JE, Kargman SL, Oshima H, Hancock B, et al.: Suppression of intestinal polyposis in $Apc^{\Delta 716}$ knockout mice by inhibition of cyclooxygenase 2 (COX-2). *Cell* **87**, 803–809, 1996.

33. Korinek V, Barker N, Morin PJ, Van Wichen DV, de Weger R, et al.: Constitutive transcriptional activation by a β-catenin-Tcf complex in *APC*−/− colon carcinoma. *Science* **275**, 1784–1787, 1997.

34. Rao CV, Cooma I, Rosa JG, Simi B, El-Bayoumy K, et al.: Chemoprevention of familial adenomatous polyposis development in the *APC^min* mouse model by 1,4-phenylenebis(methylene)selenocyanate. *Carcinogenesis* **21**, 617–621, 2000.

35. Hoffmann D and Wynder EL: Smoking and lung cancer: scientific challenges and opportunities. *Cancer Res* **54**, 5284–5295, 1994.

36. National Cancer Institute: *Smoking, Tobacco and Cancer Program 1985–1989 Status Report*. Bethesda, MD: National Institutes of Health, 1990. (NIH Publ 90-3107)

37. Perkin DM, Pisani P, and Ferlay J: Estimates of the worldwide incidence of 18 major cancers in 1985. *Int J Cancer* **54**, 594–605, 1985.

38. Hecht SS: Biochemistry, biology, and carcinogenicity of tobacco-specific N-nitrosamines. *Chem Res Toxicol* **11**, 559–603, 1998.

39. El-Bayoumy K, Upadhyaya P, Desai DH, Amin S, and Hecht SS: Inhibition of 4-(methylnitrosamino)-1-(3-pyridyl)-1-butanone tumorigenicity in mouse lung by the synthetic organoselenium compound 1,4-phenylenebis(methylene)selenocyanate. *Carcinogenesis* **14**, 1111–1113, 1993.

40. Prokopczyk B, Amin S, Desai DH, Kurtzke C, Upadhyaya P, et al.: Effects of 1,4-phenylenebis(methylene)selenocyanate and selenomethionine on 4-(methylnitrosamino)-1-(3-pyridyl)-1-butanone-induced tumorigenesis in A/J mouse lung. *Carcinogenesis* **18**, 1855–1857, 1997.

41. Prokopczyk B, Cox JE, Upadhyaya P, Amin S, Desai D, et al.: Effects of dietary 1,4-phenylenebis(methylene)selenocyanate on 4-(methylnitrosamino)-1-(3-pyridyl)-1-butanone-induced DNA adduct formation in lung and liver of A/J mice and F344 rats. *Carcinogenesis* **17**, 749–753, 1996.

42. Rosa JGV, Prokopczyk B, Desai DH, Amin SG, and El-Bayoumy K: Elevated 8-hydroxy-2'-deoxyguanosine levels in lung DNA of A/J mice and F344 rats treated with 4-(methylnitrosamino)-1-(3-pyridyl)-1-butanone and inhibition by dietary 1,4-phenylenebis(methylene)selenocyanate. *Carcinogenesis* **19**, 1783–1788, 1998.

43. Prokopczyk B, Rosa JG, Desai D, Amin S, Sohn O-S, et al.: Chemoprevention of lung tumorigenesis induced by a mixture of benzo[a]pyrene and 4-(methylnitrosamino)-1-(3-pyridyl)-1-butanone by the organoselenium compound 1,4-phenylenebis(methylene)selenocyanate. *Cancer Lett* **161**, 35–46, 2000.

44. El-Bayoumy K and Hoffmann D: Nutrition and tobacco-related cancer. In *Nutritional Oncology*, Heber D and Blackburn G (eds). San Diego, CA: Academic, 1999, pp 299–324.

45. Lippman SM and Hong WK: Second malignant tumors in head and neck squamous cell carcinoma: the overshadowing threat for patients with early-stage disease. *Int J Radiat Oncol Biol Phys* **17**, 691–694, 1989.

46. Tanaka T: Chemoprevention of oral carcinogenesis. *Eur J Cancer* **31F**, 3–15, 1995.

47. Tanaka T, Makita H, Kawabata K, Mori H, and El-Bayoumy K: 1,4-Phenylenebis(methylene)selenocyanate exerts exceptional chemopreventive activity in rat tongue carcinogenesis. *Cancer Res* **57**, 3644–3648, 1997.

48. Fidler I: The evolution of biological heterogeneity in metastatic neoplasms. In *Cancer Invasion and Metastases: Biologic and Therapeutic Aspects*, Nicolson GL and Milas L (eds). New York: Raven, 1984, pp 5–26.

49. Fidler IJ: Critical determinants of cancer metastasis: rationale for therapy. *Cancer Chemother Pharmacol* **43** Suppl, S3–S10, 1999.

50. Yan L, Yee JA, McGuire MH, and Graef GL: Effect of dietary supplementation of selenite on pulmonary metastasis of melanoma cells in mice. *Nutr Cancer* **28**, 165–169, 1997.

51. Yan L, Yee JA, Li D, McGuire MH, and Graef GL: Dietary supplementation of selenomethionine reduces metastasis of melanoma cells in mice. *Anticancer Res* **19**, 1337–1342, 1999.

52. Tanaka T, Kohno H, Murakami M, Kagami S, and El-Bayoumy K: Suppressing effects of dietary supplementation of the organoselenium 1,4-phenylenebis(methylene)selenocyanate and the citrus antioxidant auraptene on lung metastasis of melanoma cells in mice. *Cancer Res* **60**, 3713–3716, 2000.

53. Ganther HE: Selenium metabolism, selenoproteins and mechanisms of cancer prevention: complexities with thioredoxin reductase. *Carcinogenesis* **20**, 1657–1666, 1999.

54. Ip C, Thompson HJ, Zhu Z, and Ganther HE: In vitro and in vivo studies of methylseleninic acid: evidence that a monomethylated selenium metabolite is critical for cancer chemoprevention. *Cancer Res* **60**, 2882–2886, 2000.

55. El-Bayoumy K, Upadhyaya P, Date V, Sohn O-S, Fiala ES, et al.: Metabolism of [14C]benzyl selenocyanate in the F344 rat. *Chem Res Toxicol* **4**, 560–565, 1991.

56. Hasegawa T, Okuno T, Nakamuro K, and Sayato Y: Identification and metabolism of selenocysteine-glutathione selenenyl sulfide (CySeSG) in small intestine of mice orally exposed to selenocystine. *Arch Toxicol* **71**, 39–44, 1996.

57. Kawamori T, El-Bayoumy K, Ji B-Y, Rodriguez JGR, Rao CV, et al.: Evaluation of benzyl selenocyanate glutathione conjugate for potential chemopreventive properties in colon carcinogenesis. *Int J Oncol* **13**, 29–34, 1998.

58. El-Bayoumy K, Upadhyaya P, Sohn O-S, Rosa JG, and Fiala ES: Synthesis and excretion profile of 1,4-[14C]phenylenebis(methylene)selenocyanate in the rat. *Carcinogenesis* **19**, 1603–1607, 1998.

59. Lu J, Jiang C, Kaeck M, Ganther H, Vadhanavikit S, et al.: Dissociation of the genotoxic and growth inhibitory effects of selenium. *Biochem Pharmacol* **50**, 213–219, 1995.

60. Ip C, Lisk DJ, and Ganther HE: Activities of structurally related lipophilic selenium compounds as cancer chemopreventive agents. *Anticancer Res* **18**, 4019–4026, 1998.

NUTRITION AND CANCER, *40*(1), 28–33
Copyright © 2001, Lawrence Erlbaum Associates, Inc.

Mechanisms of Organoselenium Compounds in Chemoprevention: Effects on Transcription Factor-DNA Binding

B. Woo Youn, Emerich S. Fiala, and Ock Soon Sohn

Abstract: Data obtained on the effects of selenium compounds on regulatory transcription factor-DNA binding by other laboratories are briefly reviewed, and some of our own results in this area are also presented. We assessed the in vitro and in vivo effects of the organoselenium compound 1,4-phenylenebis(methylene)selenocyanate (p-XSC) on the binding activities of the transcription factors nuclear factor-κB (NF-κB), activator protein-1 (AP-1), Sp1, and Sp3 using the HCT-116 (human colorectal adenocarcinoma) cell line as a model system. Using nuclear extracts, electrophoretic mobility shift assays were carried out to determine the extent of binding of the transcription factors to their respective consensus recognition sites on radiolabeled oligonucleotides. p-XSC and sodium selenite reduced the consensus site binding activity of NF-κB in a concentration-dependent manner when nuclear extracts from cells stimulated with tumor necrosis factor-α were incubated with either compound ("in vitro"). However, only p-XSC inhibited NF-κB consensus recognition site binding when the cells were pretreated with either compound and were then stimulated with tumor necrosis factor-α ("in vivo"). In contrast, the consensus site binding activity of AP-1 was inhibited only with sodium selenite, but not with p-XSC in vitro or in vivo. p-XSC or sodium selenite reduced the consensus site binding of transcription factors Sp1 and Sp3 in concentration- and time-dependent manners when nuclear extracts from cells treated with either compound in vivo were assayed by electrophoretic mobility shift assay. 1,4-Phenylenebis(methylene)thiocyanate, the sulfur analog of p-XSC, which is inactive in chemoprevention, had no effect on the oligonucleotide binding of Sp1 and Sp3. Our observations could provide further clues as to the mechanisms involved in the chemoprevention of cancer by p-XSC.

Introduction

There is excellent evidence for the efficacy of certain inorganic and organic forms of selenium as cancer-chemopreventive compounds in humans (1–3) and in animal models (4–6). With the conviction that the discovery and rational development of even more effective forms of selenium chemopreventive agents depends on a knowledge of their mechanisms of action, we undertook studies to examine the possible effects of the organoselenium chemopreventive compounds developed by El-Bayoumy et al. (4) on cellular regulatory transcription factors. Regulatory transcription factors (referred to here simply as "transcription factors") are proteins that serve to modulate the level of transcription of DNA to mRNA by binding to specific base sequences on gene promoters. These proteins directly or indirectly influence every aspect of normal cellular activities, including cell proliferation, differentiation, and apoptosis. Their expression or activation may be either constitutive, meeting basic intracellular demands, or inducible by a broad range of external stimuli. A permanent loss or alteration in the activities of transcription factors through deletion or mutation can result in the deregulation of signal transduction pathways that normally respond to the needs of the organism as well as to cellular stress and cell damage. Such loss/alterations can lead to cell death or oncogenesis. On the other hand, a nonhereditary, temporary loss of function, such as might be caused by pharmacological agents, including selenium compounds, could lead to cell death. If precancerous or cancerous cells were more sensitive to these agents than normal cells (a possible reason for this is presented below), the result would be chemoprevention or the inhibition of carcinogenesis. Studies by others on the alteration of nuclear factor-κB (NF-κB), activator protein-1 (AP-1), Sp1, and Sp3 transcription factor activities by selenium compounds and a description of some of our own recent research in this area are the subjects of this brief review.

Our laboratory is examining the in vitro and in vivo effects of the organoselenium compound 1,4-phenylene-bis(methylene)selenocyanate (*p*-XSC) (4) on the activities of several transcription factors, including NF-κB, AP-1, Sp1, and Sp3, using the HCT-116 (human colorectal adenocarcinoma) cell line as a model system. "In vitro" effects are examined by adding the selenium compound to nuclear extracts prepared from cells that had been stimulated by tumor necrosis factor-α (TNF-α). "In vivo" effects are studied by using nuclear extracts from cells pretreated with the compound, with or without TNF-α stimulation. In either case, the nuclear extracts are assayed by the electrophoretic mo-

The authors are affiliated with the American Health Foundation, Valhalla, NY 10595.

bility shift assay (EMSA) for binding to oligonucleotides that incorporate the transcription factor consensus recognition sequence. Here we provide evidence that NF-κB activation is inhibited by *p*-XSC in vivo and that oligonucleotide binding by Sp1 and Sp3 is reduced in a concentration-dependent manner by *p*-XSC in vivo. Moreover, 1,4-phenylenebis(methylene)thiocyanate (*p*-XTC), the sulfur analog of *p*-XSC, which is inactive in chemoprevention, does not inhibit the oligonucleotide binding of Sp1 or Sp3, emphasizing the essential role of selenium in this effect. These observations provide further information on the complex mechanisms involved in *p*-XSC cancer chemoprevention.

p-XSC Inhibition of NF-κB Consensus Sequence Binding

The most abundant and most studied member of the NF-κB transcription factor family, the heterodimer composed of p65 and p50 subunits, is involved in the regulation of a large number of genes that control various aspects of the immune and stress responses, inflammation, and apoptosis (reviewed in Ref. 7). Activation of NF-κB, which is normally present in the cytoplasm, involves the dissociation of an inhibitor, IκB, and translocation of the transcription factor to the nucleus with consequent binding to specific sequences on the promoters of genes, modulating (generally increasing) their transcription. There is increasing evidence that, in many cell types, the activation of NF-κB leads to the inhibition of apoptosis and that the deregulation of NF-κB activation, in many cases, is associated with tumorigenesis (7); the overexpression and amplification of the genes coding for NF-κB commonly occur in a large number of human tumors (8).

The activation of NF-κB can be inhibited by many different agents (9), and we call attention to the fact that most of these NF-κB inhibitors are known to be effective in the chemoprevention or therapy of cancer. To us, these considerations suggest a strong causal relationship between the inhibition of NF-κB and the inhibition of carcinogenesis.

The inhibition of NF-κB activation by selenium compounds, with consequent repression of certain of its target genes, has been demonstrated previously by others. It has been reported that supplementation of selenium-deprived Jurkat and Esb-L T lymphocytes with sodium selenite inhibited NF-κB activation in response to TNF-α treatment (10). The same work used reporter gene assays with luciferase constructs to demonstrate a dose-dependent selenite inhibition of NF-κB-controlled HIV-1 long terminal repeat expression in Jurkat cells. In a similar study, the DNA (i.e., oligonucleotide) binding ability of NF-κB was significantly inhibited in nuclear extracts of Jurkat T cells or human lung adenocarcinoma cells after in vitro or in vivo treatment with sodium selenite (11). In the latter work, in vivo treatment of both cell types with selenite also inhibited the induction, by bacterial lipopolysaccharide, of the inducible form of nitric oxide synthase, a product of one of the genes controlled by NF-κB. The in vitro inhibition of DNA binding was reversed by the addition of the reducing agent dithiothreitol (DTT), indicating that the formation of DTT-reducible selenium-sulfur bonds was probably involved in the effect. In neither of these studies, however, was a connection made between the inhibition of NF-κB binding by selenite and its cancer-chemopreventive capability.

Using EMSA, we examined the in vitro and in vivo effects of *p*-XSC (Fig. 1) on the ability of NF-κB to bind to its

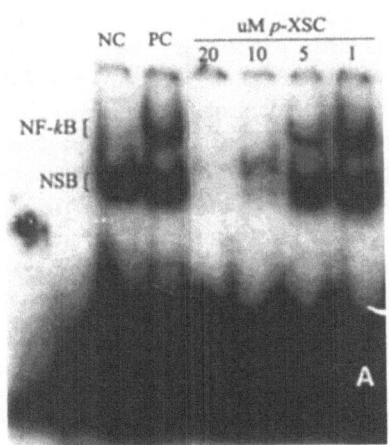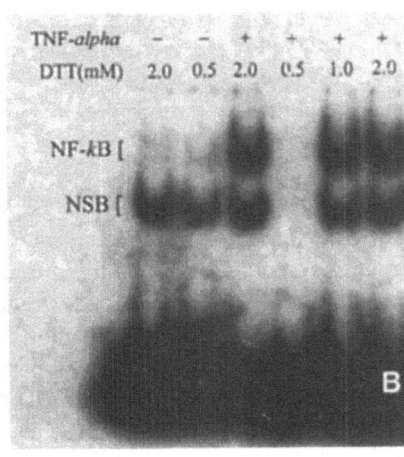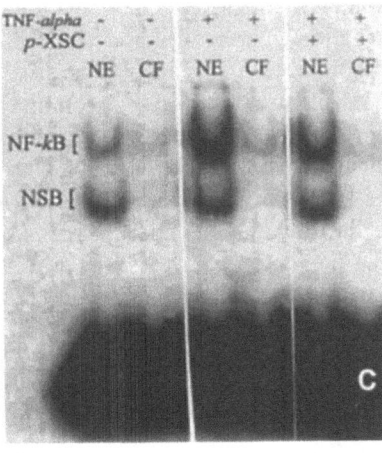

Figure 1. In vitro and in vivo effects of 1,4-phenylenebis(methylene)selenocyanate (*p*-XSC) on binding of nuclear factor-κB (NF-κB) to its consensus recognition sequence. A: dose-dependent inhibition of NF-κB-oligonucleotide binding in vitro. HCT-116 cells were stimulated with tumor necrosis factor-α (TNF-α) (2 ng/ml) for 30 min. Nuclear extract was prepared and analyzed by electrophoretic mobility shift assay (EMSA) in the presence of various concentrations of *p*-XSC. Complete inhibition of oligonucleotide binding by *p*-XSC was observed at 10 μM *p*-XSC; partial inhibition was seen at 5 μM. NC, negative control; PC, positive control; NSB, nonspecific bands. B: reversal by dithiothreitol (DTT) of inhibition of NF-κB recognition sequence binding in vitro. HCT-116 cells were treated with TNF-α (2 ng/ml) for 30 min, and nuclear extract was prepared and analyzed by EMSA in the presence of 10 μM *p*-XSC with subsequent addition of 0.5, 1.0, or 2.0 mM DTT. Complete reversal of inhibition of binding occurred as DTT concentration was increased from 0.5 to 1.0 mM. C: effect of *p*-XSC on binding property of NF-κB in vivo using nuclear extract and cytoplasmic fraction. HCT-116 cells were pretreated with 20 μM *p*-XSC for 1 h before they were stimulated with TNF-α (2 ng/ml) for 30 min. Nuclear extract (NE) and cytoplasmic fraction (CF) were prepared and analyzed by EMSA. Inhibition was detected in nuclear extract only.

oligonucleotide consensus sequence. The results in Fig. 1A show that p-XSC inhibits the binding of NF-κB in vitro and that the inhibition is concentration dependent, with complete inhibition occurring at 10 μM. We also observed that the inhibition was reversed in the presence of 1 mM DTT (Fig. 1B), suggesting that the formation of –Se-S- bonds between p-XSC and essential –SH groups of the transcription factor may have been responsible for the inhibition, in analogy to the proposed action of selenite (11). To find out whether p-XSC had an analogous effect in vivo, cells were treated with the compound for 1 h, the medium was changed, and the cells were stimulated with TNF-α for 30 min. Nuclear and the cytoplasmic extracts were prepared and used for EMSA. p-XSC inhibited the binding of NF-κB with the nuclear extracts; the cytoplasmic extracts by themselves showed no binding activity, suggesting that the effect of p-XSC was not due merely to the inhibition of the translocation of NF-κB from the cytoplasm to the nucleus (Fig. 1C).

Consensus Recognition Sequence Binding Activity of AP-1 in p-XSC-Treated Cells

AP-1 transcription factors, such as the Jun-Jun homodimer or the Jun-Fos heterodimer, bind to 12-O-tetradecanoylphorbol-13-acetate response elements (TRE), transcriptionally activating a variety of effector genes. The c-jun and c-fos genes are inducible by a broad range of extracellular stimuli and function as intermediary transcriptional regulators in signal transduction processes leading to cell proliferation and transformation (12). The role(s) of AP-1 in cell apoptosis remains unclear. AP-1 can modulate stress-induced apoptosis positively or negatively, depending on the microenvironment and the cell type in which the stress is induced (13). The binding of AP-1 to its DNA recognition site was shown to be inhibited by selenodiglutathione and sodium selenite in nuclear extracts of 3B6 lymphocytes, and selenite also inhibited AP-1 activation in vivo (14). As in previous studies on NF-κB (11) mentioned above, the inhibition of AP-1-DNA binding by selenite in vitro was prevented by the presence of the reducing agent DTT. In this case, the authors suggested that the inhibition of AP-1 binding could be important in the chemopreventive effects of selenium. In another study (15), sodium selenite was also reported to inhibit AP-1 binding in vitro, as well as to repress the expression of an AP-1-dependent reporter gene in MCF-7 cells. Although in preliminary studies we observed an inhibition of AP-1-oligonucleotide binding by sodium selenite, we were unable to detect an inhibition of binding by p-XSC in vitro, or in vivo after TNF-α stimulation of HCT-116 cells (data not shown). This difference between the effects of selenite and p-XSC suggests the occurrence of interesting qualitative or quantitative differences in the binding of the two selenium compounds to the subunit components of AP-1.

Inhibition of Sp1 and Sp3 Consensus Recognition Sequence Binding by p-XSC

The Sp1 and Sp3 transcription factors are ubiquitously expressed at high levels in many mammalian cell lines (reviewed in Ref. 16). These transcription factors recognize the guanine- and cytosine-rich oligonucleotide sequence known as the "G/C box." Sp3 is known to play antagonistic (17) as well as synergistic roles in the regulation of gene expression by Sp1, depending on the gene, the cell type, and the cellular context. Sp1 sites often control initiation of transcription from downstream sites in TATA-less promoters, which are prevalent among late-G_1 genes. Inhibition of the transactivational activity of Sp1 could strongly impact cell proliferation, since several of the Sp1 target gene products, such as dihydrofolate reductase, ornithine decarboxylase, and thymidine kinase play important roles in replication of DNA (17,18). Others, such as p21[CIP1/WAF1] (19), are involved in cell cycle regulation, cell differentiation, and apoptosis. The members of the Sp family of transcription factors probably interact with each other to assist in adequate and timely expression of target genes during the cell cycle and on extracellular stimulation. It appears, however, that only a single member of this family, i.e., Sp1, plays a predominant role in the control of gene expression. Interestingly, certain genes involved in the inhibition of apoptosis, namely, survivin and bcl-2, also contain Sp1 binding sites in their promoter regions (20,21). Thus it is likely that Sp1 plays an important role not only in the regulation of cell growth and proliferation but also in programmed cell death.

Other observations imply a wider role for the Sp1 transcription factor in DNA cytosine methylation-mediated epigenetic control of gene expression. Thus Sp1 binding sites have been shown to be critical for the maintenance of the methylation-free CpG island of the aprt gene (22,23), and it is likely that the methylation status of the CpG islands of other genes is under similar control. Because Sp1 binding sites occur very frequently in CpG islands, this suggests that occupied Sp1 sites prevent the methylation-induced silencing of many CpG island-containing genes (24,25).

To our knowledge, effects of selenium compounds on the Sp family of transcription factors have not been reported. In preliminary studies, we examined the dose dependency and time course of p-XSC effects on the consensus recognition site binding properties of Sp1 and Sp3 in nuclear extracts of HCT-116 cells. A dose-dependent decrease in Sp1 and Sp3 oligonucleotide binding was observed on the addition of the organoselenium compound to the EMSA incubation mixture in vitro (Fig. 2) as well as in vivo, when HCT-116 cells were treated with various concentrations of p-XSC for 24 h (Fig. 3). The inhibition in Sp1 and Sp3 binding appeared to increase with increasing time of exposure of the cells to the organoselenium compound (Fig. 4). Antibody supershift experiments (results not shown) demonstrated that the first slow-migrating band on the gel in Fig. 4 contains Sp1-oligonucleotide and Sp3-oligonucleotide complexes and the

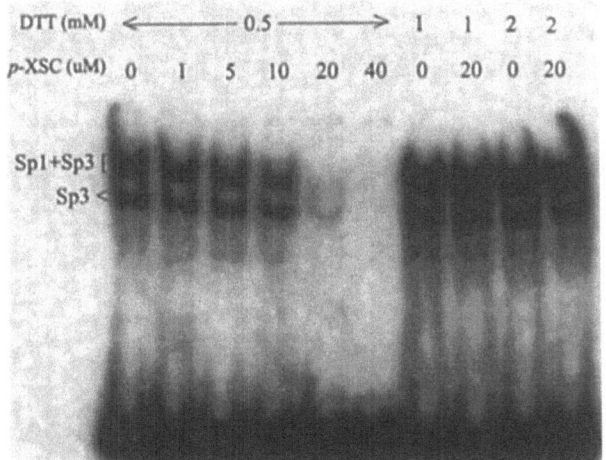

Figure 2. Dose-dependent in vitro inhibition of Sp1-oligonucleotide binding by *p*-XSC and its reversal by DTT. HCT-116 cells were stimulated with TNF-α (2 ng/ml) for 30 min. Nuclear extract was prepared and analyzed by EMSA in the presence of various concentrations of *p*-XSC. Complete inhibition of binding by *p*-XSC was observed at ≥20 μM. EMSA incubations in the presence of 20 μM *p*-XSC with subsequent addition of 0.5, 1.0, or 2.0 mM DTT showed complete reversal of inhibition as DTT concentration was increased from 0.5 to 1.0 mM.

second fast-migrating band contains the Sp3-oligonucleotide complex. In related studies, we treated the cells with various concentrations of *p*-XTC for 24 h, but essentially no changes in Sp1 and Sp3 binding were noted at any of the *p*-XTC concentrations used. We performed analogous experiments using sodium selenite. As in the case of *p*-XSC, a similar dose-dependent decrease in Sp1 and Sp3 oligonucleotide binding in nuclear extracts was observed when the cells were treated with various concentrations of sodium selenite for 24 h (Fig. 3). Time-course experiments in Fig. 4 showed that Sp1 and Sp3 binding disappeared rapidly when the cells were treated with 10 μM sodium selenite for different lengths of time.

Discussion

In these initial studies, we have used EMSA to show that the organoselenium chemopreventive compound *p*-XSC inhibits the activities of important transcription factors in vivo, when HCT-116 cells are treated with the compound, and in vitro, when *p*-XSC is added directly to the EMSA incubation mixture. It remains to be demonstrated that the observed inhibition of binding is associated with decreased expression of genes specifically under control of NF-κB or Sp1/Sp3 and that this occurs not only in cell cultures but also in animal tissue. For the ensuing discussion, we assume that this will indeed be the case.

With regard to the inhibition of NF-κB binding, previous reports have ascribed sodium selenite- and *p*-XSC-induced lethal toxicity in a mouse mammary carcinoma cell line (26) and in human colon carcinoma HT-29 cells (27) to the induction of apoptosis. In view of the ability of sodium selenite (10,11) and *p*-XSC (results presented here) to inhibit the binding of NF-κB to its DNA recognition sequence, we suggest that the inhibition of NF-κB binding may, in part, be responsible for the observed apoptotic effect. According to Barkett and Gilmore (7), there are three general models by which the NF-κB transcription factor may regulate apoptosis: 1) direct regulation of genes that inhibit (7,28,29) or promote apoptosis, 2) regulation of the cell cycle, which then desensitizes a cell to apoptotic signals, and 3) interaction with cellular proteins, for example, cotranscription factors, such as CBP/p300, the levels of which affect the cross talk/interplay between transcription factors, as shown in the case of NF-κB and p53 (30). In terms of the direct regulation of genes, for example, the inhibition of NF-κB activation and consequent loss of gene transactivation caused by exposure to *p*-XSC in vivo could lead to the downregulation of inhibitors of apoptosis (IAP), one well-known group of NF-κB target genes (9,29), resulting in apoptosis, as in fact observed (26,

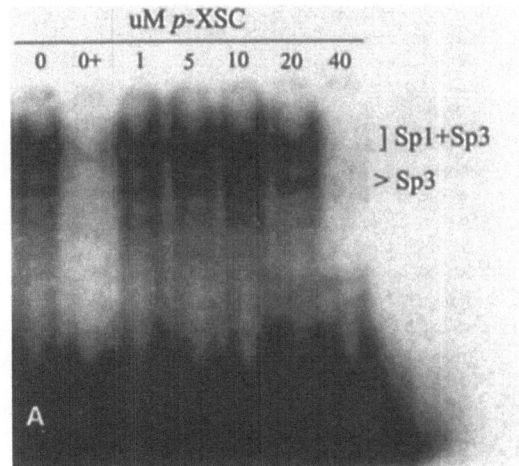

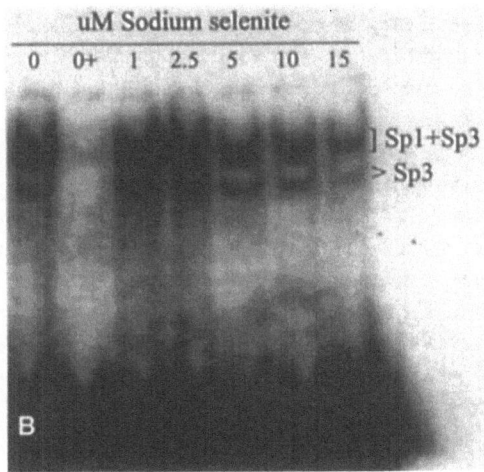

Figure 3. Dose-dependent effect of *p*-XSC (A) and sodium selenite (B) on consensus recognition sequence binding of Sp1 and Sp3 in vivo. HCT-116 cells were treated with various concentrations of *p*-XSC or sodium selenite for 24 h. Nuclear extracts were prepared and analyzed by EMSA. Compared with controls, both selenium compounds caused Sp1 and Sp3 oligonucleotide binding to decrease. 0+ lanes, EMSA performed after addition of excess unlabeled oligonucleotide containing consensus recognition sequence.

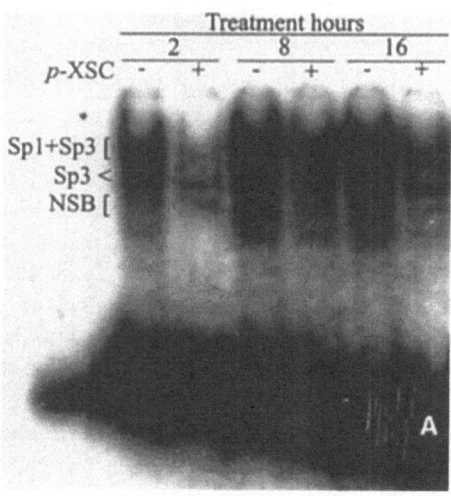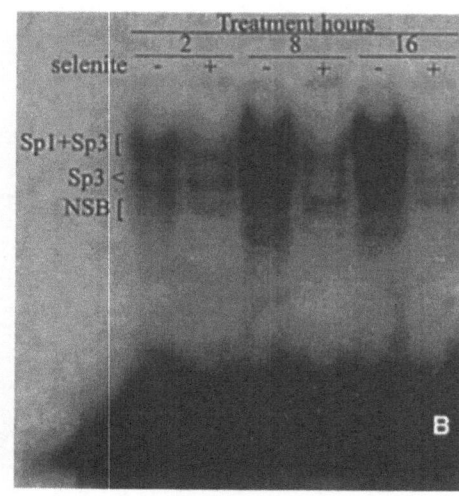

Figure 4. Time course of effects of *p*-XSC (A) and sodium selenite (B) on binding property of Sp1 and Sp3 in vivo. HCT-116 cells were treated with 10 μM *p*-XSC for various lengths of time. Nuclear extracts were prepared and analyzed by EMSA for oligonucleotide binding of Sp1 and Sp3. Increasing inhibition of binding occurred with increasing time of *p*-XSC treatment.

27). Whether *p*-XSC induces apoptosis in the HCT-116 cell line, the specific role of NF-κB in the effect and the mechanism of its inhibition are questions that we are examining.

As previously mentioned, Sp1 binding sites are present in the promoters of the antiapoptotic genes *bcl-2* (21) and *survivin* (20), being especially numerous in the latter. Both of these genes are overexpressed in many human tumors. Thus, as in the case of NF-κB, the prevention of Sp1 binding to its recognition sites on these genes by inorganic or organic selenium, as shown in this work, might lead to the preferential induction of apoptosis in those cells that overexpress antiapoptotic genes. This mechanism might offer a simple explanation for the (presumed—likely, but not as yet demonstrated) greater sensitivity of cancerous and precancerous cells to the toxicity of chemopreventive selenium compounds. In addition to proapoptotic effects, inhibition of NF-κB and Sp1 binding to DNA may also be expected to reduce the transcriptional activity of genes coding for proteins necessary for, or accessory to, cell proliferation. In the case of NF-κB, these might include protooncogenes such as c-*myc* (31) and, in the case of Sp1, enzymes including dihydrofolate reductase and ornithine decarboxylase (16–18). Thus we suggest that the simultaneous repression of antiapoptotic and proproliferative genes may be responsible for the high chemopreventive efficacy of *p*-XSC and sodium selenite. Further studies are also necessary to precisely localize the sites, presumably cysteine residues, on the NF-κB and Sp1/Sp3 transcription factors, that are modified by the selenium compounds and to characterize the modifications, since the results could lead to the development of even more effective selenium cancer-chemopreventive compounds.

Acknowledgments and Notes

This work is dedicated to the memory of Dr. Larry C. Clark, an outstanding and pioneering scientist in the field of human cancer chemoprevention and a wonderful human being. This work was supported by National Cancer Institute Grant CA-46589. Address correspondence to Emerich S. Fiala, American Health Foundation, 1 Dana Rd., Valhalla, NY 10595.

Submitted 15 November 2000; accepted in final form 30 November 2000.

References

1. Clark LC, Dalkin B, Krongrad A, Combs GR Jr, Turnbull BW, et al.: Decreased incidence of prostate cancer with selenium supplementation: results of a double-blind cancer prevention trial. *Br J Urol* **81**, 730–734, 1998.
2. Clark LC, Combs GR Jr, Turnbull BW, Slate EH, Chalker DK, et al.: Effects of selenium supplementation for cancer prevention in patients with carcinoma of the skin: a randomized controlled trial. *JAMA* **276**, 1957–1963, 1996.
3. Combs GF Jr, Clark LC, and Turnbull BW: Reduction of cancer mortality and incidence by selenium supplementation. *Med Klin* **92** *Suppl 3*, 42–45, 1997.
4. El-Bayoumy K, Upadhyaya P, Chae Y-H, Sohn OS, Rao CV, et al.: Chemoprevention of cancer by organoselenium compounds. *J Cell Biochem Suppl* **22**, 92–100, 1995.
5. Ip C: Lessons from basic research in selenium and cancer prevention. *J Nutr* **128**, 1845–1854, 1998.
6. Reddy BS, Rivenson A, Kulkarni N, Upadhyaya P, and El-Bayoumy K: Chemoprevention of colon carcinogenesis by the synthetic organoselenium compound 1,4-phenylenebis(methylene)selenocyanate. *Cancer Res* **52**, 5635–5640, 1992.
7. Barkett M and Gilmore TD: Control of apoptosis by Rel/NF-κB transcription factors. *Oncogene* **18**, 6910–6924, 1999.
8. Rayet B and Gelinas C: Aberrant *rel/nfkb* genes and activity in human tumors. *Oncogene* **18**, 6938–6947, 1999.
9. Epinat J-C and Gilmore TD: Diverse agents act at multiple levels to inhibit the Rel/NF-κB signal transduction pathway. *Oncogene* **18**, 6896–6909, 1999.
10. Makropoulos V, Bruning T, and Schulze-Osthoff K: Selenium-mediated inhibition of transcription factor NF-κB and HIV-1 LTR promotor activity. *Arch Toxicol* **70**, 277–283, 1996.
11. Kim IY and Stadtman TC: Inhibition of NF-κB DNA binding and nitric oxide induction in human T cells and lung adenocarcinoma cells by selenite treatment. *Proc Natl Acad Sci USA* **94**, 12904–12907, 1997.
12. Sun Y and Oberley LW: Redox regulation of transcriptional activators. *Free Radic Biol Med* **21**, 335–348, 1996.

13. Libermann DA, Gregory B, and Hoffman B: AP-1 (Fos/Jun) transcription factors in hematopoietic differentiation and apoptosis. *Int J Oncol* **12**, 685–700, 1998.

14. Spyrou G, Björnstedt M, Kumar S, and Holmgren A: AP-1 DNA-binding activity is inhibited by selenite and selenodiglutathione. *FEBS Lett* **368**, 59–63, 1995.

15. Handel ML, Watts CK, DeFazio A, Day RO, and Sutherland RL: Inhibition of AP-1 binding and transcription by gold and selenium involving conserved cysteine residues in Jun and Fos. *Proc Natl Acad Sci USA* **92**, 4497–4501, 1995.

16. Phillipsen S and Suske G: A tale of three fingers: the family of mammalian Sp/XKLF transcription factors. *Nucleic Acids Res* **27**, 2991–3000, 1999.

17. Kumar AP and Butler AP: Transcription factor Sp3 antagonizes activation of the ornithine decarboxylase promoter by Sp1. *Nucleic Acids Res* **25**, 2012–2019, 1997.

18. Lania L, Majello B, and De Luca P: Transcriptional regulation by the Sp family members. *Int J Cell Biol* **29**, 1313–1323, 1997.

19. Moustakas A and Kardassis D: Regulation of the human p21[Waf1/Cip1] promoter in hepatic cells by functional interactions between Sp1 and Smad family members. *Proc Natl Acad Sci USA* **95**, 6733–6738, 1998.

20. Li F and Altieri DC: Transcriptional analysis of human *survivin* gene expression. *Biochem J* **344**, 305–311, 1999.

21. Dong L, Wang W, Wang F, Stoner M, Reed JC, et al.: Mechanisms of transcriptional activation of *bcl-2* gene expression by 17β-estradiol in breast cancer cells. *J Biol Chem* **274**, 32099–32107, 1999.

22. Brandels M, Frank D, Keshet I, Siegfried Z, Mendelsohn M, et al.: Sp1 elements protect a CpG island from de novo methylation. *Nature* **371**, 435–438, 1994.

23. Macleod D, Charlton J, Mullins J, and Bird A: Sp1 sites in the mouse *aprt* gene promotor are required to prevent methylation of the CpG island. *Genes Dev* **8**, 2282–2292, 1994.

24. Marin M, Karis A, Visser P, Grosveld F, and Philipsen S: Trancription factor Sp1 is essential for early embryonic development but dispensable for cell growth and differentiation. *Cell* **89**, 619–628, 1997.

25. Mancini DN, Singh SM, Archer TK, and Rodenhiser DI: Site-specific DNA methylation in the neurofibromatosis (NF1) promotor with binding of CREB and SP1 transcription factors. *Oncogene* **18**, 4108–4119, 1999.

26. Thompson HJ, Wilson A, Lu J, Singh M, Jiang C, et al.: Comparison of the effects of an organic and an inorganic form of selenium on a mammary carcinoma cell line. *Carcinogenesis* **15**, 183–186, 1994.

27. Stewart MS, Davis RL, Walsh LP, and Pence BC: Induction of differentiation and apoptosis by sodium selenite in human colonic carcinoma cells (HT-29). *Cancer Lett* **117**, 35–40, 1997.

28. Wang CY, Mayo MW, Korneluk RG, Goeddel DV, and Baldwin AS Jr: NF-κB antiapoptosis: induction of TRAF1 and TRAF2 and c-IAP1 and c-IAP2 to suppress caspase-8 activation. *Science* **281**, 1680–1683, 1998.

29. LaCasse EC, Baird S, Korneluk RG, and MacKenzie AE: The inhibitors of apoptosis (IAPs) and their emerging role in cancer. *Oncogene* **17**, 3247–3259, 1998.

30. Webster GA and Perkins ND: Transcriptional cross talk between NF-κB and p53. *Mol Cell Biol* **19**, 3485–3495, 1999.

31. Pahl HL: Activators and target genes of Rel/NF-κB transcription factors. *Oncogene* **18**, 6853–6866, 1999.

NUTRITION AND CANCER, *40*(1), 34–41

Dimethyldiselenide and Methylseleninic Acid Generate Superoxide in an In Vitro Chemiluminescence Assay in the Presence of Glutathione: Implications for the Anticarcinogenic Activity of L-Selenomethionine and L-Se-Methylselenocysteine

Julian E. Spallholz, Brent J. Shriver, and Ted W. Reid

Abstract: The reduction of cancer incidence by dietary supplementation with L-selenomethionine, L-Se-methylselenocysteine, and other methylated selenium compounds and metabolites is believed to be due to the metabolic generation of the monomethylated selenium species methylselenol. Dimethyldiselenide and methylseleninic acid were reduced by glutathione in an in vitro chemiluminescent assay in the presence of lucigenin for the detection of superoxide ($O_2^-\cdot$). The methylselenol produced on reduction of dimethyldiselenide and methylseleninic acid was found to be highly catalytic, continuously generating a steady state of $O_2^-\cdot$. The $O_2^-\cdot$ detected by the chemiluminescence generated by methylselenol was fully quenched by superoxide dismutase, causing a complete cessation of chemiluminescence. In contrast, dimethyldisulfide in the presence of glutathione was not catalytic to any measurable extent and did not generate any superoxide. These in vitro results suggest that methylselenol catalysis is possible in vivo, and if metabolism generates sufficient concentrations of methlylselenol from L-selenomethionine or L-Se-methylselenocysteine in vivo, it could change the redox status of cells and oxidatively induce cellular apoptosis.

Introduction

Selenium is an essential nutrient for humans; it fulfills the physiological requirements for ≥13 human enzymes and proteins, the biological functions of which have been partially or fully characterized (1). The first of these human proteins to be characterized were the glutathione peroxidases, which account for selenium's antioxidant role in the reduction of H_2O_2 to water and organic hydroperoxides to alcohols. The form of selenium in all selenoproteins is an amino acid, L-selenocysteine (2). The nutritional requirement for selenium for the synthesis of L-selenocysteine is fulfilled by the consumption of cereal grains, meats, and seafood (3). The present selenium US Recommended Dietary Allowance for men and women is 55 μg/day, being recently reduced from 70 μg/day for men (4).

Selenium consumed as a dietary supplement, beyond that found in foods, at up to 200 μg/day, has been reported by Clark et al. (5) to reduce the incidence of lung, colorectal, and prostate cancer in humans. This clinical trial, reported in 1996, followed the many reports that supplements of dietary selenium in animals beyond that required to saturate all selenium proteins and enzymes prevented and/or reduced cancer. In animals, selenium supplementation reduced the size and incidence of transplanted cancers and reduced the incidence of spontaneous cancer and cancer induced by carcinogens (6–10). Thus the 1996 clinical intervention data of Clark et al. showing a reduction in cancer incidence agrees with the earlier published animal data showing the reduction in cancer by selenium at supraphysiological levels.

The metabolism of ingested selenium in animals and humans is well known (Fig. 1) (11). In general, all organic selenoamino acids found in the diet and selenite and selenate found in some dietary supplements are believed to be ultimately reduced to hydrogen selenide (H_2Se). From H_2Se the selenium may be phosphorylated and incorporated into proteins as L-selenocysteine, formed from the cotranslational modification of seryl-tRNA (12). Alternatively, under normal dietary selenium intake and during periods of excessive levels of dietary selenium ingestion, H_2Se is rapidly methylated, initially to methylselenol (CH_3SeH), then to dimethyselenide, and finally to the major excretory form of selenium found in urine, the trimethylselenonium ion. Because H_2Se is considered a very toxic form of selenium, methylation is seen to be the metabolic pathway of selenium detoxification for urinary excretion (13). Under conditions of toxic levels of selenium ingestion, this methylation metabolic pathway becomes rate limiting, and dimethylselenide is eliminated by pulmonary excretion.

Ip, Ganther, and their colleagues (14–16) have been prominent in elucidating the organic chemical forms of sele-

The authors are affiliated with Food and Nutrition, Texas Tech University, and the Department of Ophthalmology, Texas Tech University Health Sciences Center, Lubbock 79409; and Selenium Technologies, Inc., Lubbock, TX 79413.

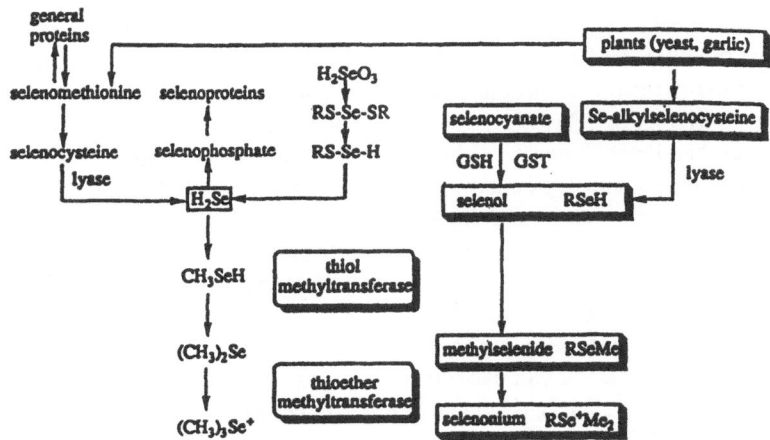

Figure 1. Selenium metabolism, emphasizing reactions for generating possible chemopreventative metabolites. GSH, glutathione; GST, glutathione S-transferase. Reprinted from *Tetrahedron*, **53**, HE Ganther and JR Lawrence, Chemical transformations of selenium in living organisms: Improved forms of selenium for cancer prevention, 12229–12310, Copyright 1997, with permission from Elsevier Science.

nium that inhibit and those that do not inhibit mammary cancer in animals. They have demonstrated a reduction of mammary cancer using the 7,12-dimethylbenz[*a*]anthracene (DMBA) chemically induced mammary cancer model by supplementing the diets of rats with the various metabolites of the selenium methylation pathway. The results of several metabolic studies have shown that the generation of H_2Se is not necessary for the carcinostatic activity of selenium, but rather it is the continuous generation of the monomethylated selenium species that imparts selenium's anticancer activity. Thus it has been determined that generation of the monomethylated selenium species CH_3SeH impairs the initiation and/or development of cancer in the DMBA mammary cancer model (16). Ganther, in a review on the chemoprevention of cancer by CH_3SeH in 1999 (17), suggests that this selenium metabolic species may oxidize thiol compounds, such as cysteine residues of proteins and/or enzymes in a catalytic role, thereby transforming the redox environment of cells. A more recent publication suggests that selenium's anticarcinogenic activity may be to reduce the blood supply

to tumors by restriction of capillary vessel development (18).

Spallholz (19,20) suggested in 1994 that the anticarcinogenic property of selenium compounds is likely due to the known toxicity of selenium compounds as studied in animals and humans. His arguments are supported by the facts that 1) selenium compounds have anticarcinogenic activity only when consumed in supraphysiological amounts above normal dietary selenium levels, i.e., >1–2.0 µg/g diet, and 2) there are different thresholds of carcinostatic activity of selenium compounds in vitro or in vivo that are highly dependent on the chemical form of selenium. Differences in the toxicity of selenium compounds in vitro and in vivo can be shown to be attributable, in part, to selenium's ability to generate superoxide ($O_2^-\cdot$) in vitro in the presence of reduced glutathione (GSH) (Table 1). Spallholz (19,20), Yan and Spallholz (21), and Xu et al. (22) also showed that all selenium compounds that have been tested that can be readily reduced by GSH to form the selenoate anion (RSe^-) are capable of generating chemiluminescence (CL) indicative of

Table 1. Selenium Compounds That Generate and Do Not Generate Superoxide In Vitro[a]

Superoxide Generated In Vitro	Superoxide Not Generated In Vitro
Selenite	Elemental selenium
Selenium dioxide	Selenate
Selenocystine	Selenomethionine
Selenocystamine	Se-methylselenocysteine[b]
Diselenopropionic acid	Selenobetaine[b]
Diphenyldiselenide	Dimethylselenoxide[b]
Dibenzyldiselenide	Selenopyridine[e]
1,4-Phenylenebis(methylene)selenocyanate[c]	Triphenylselenonium ion[b]
6-Propylselenouricil[d]	K-selenocyanate
Dimethyldiselenide	
Methylseleninic acid[b]	

a: Assay conditions are as follows: 0.05 M borate buffer, pH 9.2, containing glutathione (4 mg/ml) and lucigenin (20 µg/ml) at 25°C.
b: Courtesy of Dr. Howard Ganther (University of Wisconsin, Madison, WI).
c: Courtesy of Dr. Karam El-Bayoumy (American Health Foundation, Valhalla, NY).
d: Courtesy of Dr. Alvin Taurog (University of Texas Southwestern Medical Center, Houston, TX).
e: Courtesy of Dr. Ahmad Khalil (Yamouk University, Irbid, Jordan).

O_2^- generation in an in vitro chemical assay. In contrast, selenium compounds that do not readily form RSe⁻ by GSH reduction do not produce O_2^- in the in vitro CL assay and are much less toxic ex vivo and in vivo.

Because L-selenomethionine has been shown to reduce the incidence of cancer in humans (5) and L-Se-methyselenocysteine has been shown to be carcinostatic in animals (16) and because both compounds are theoretically capable of metabolically generating CH_3SeH in vivo, we sought to ascertain whether CH_3SeH, the reduction product of dimethyldiselenide ($CH_3SeSeCH_3$) or methlyseleninic acid (CH_3SeOOH), would generate O_2^- in vitro. For comparison, we also tested the ability of methylthiol (CH_3SH), the reduction product of dimethyldisulfide (CH_3SSCH_3), to generate O_2^- in vitro in the same CL assay.

Materials and Methods

$CH_3SeSeCH_3$ and CH_3SSCH_3 were purchased from Aldrich Chemical (Milwaukee, WI) and Sigma Chemical (St. Louis, MO), respectively. CH_3SeOOH was a gift from Dr. Howard Ganther (University of Wisconsin, Madison, WI). With use of a micropipette, aliquots of these compounds were added directly to 1.0 ml of a buffered cocktail solution comprised of sodium borate or sodium phosphate (0.05 M), lucigenin (20 μg/ml), and reduced glutathione (GSH, 4 mg/ml; Sigma Chemical). Two different buffered cocktail solutions were employed, one at pH 9.2 (borate), which is optimum for the generation, detection, and quenching of selenium-catalyzed O_2^-, and one at pH 7.4 (phosphate), which is reflective of physiological conditions. One milliliter of the cocktail solution was added to a 10 × 50-mm polypropylene tube used for the CL assay. CL using lucigenin as the detector of O_2^- (23) was counted in repetitive integrated 30-s increments over time using a chemiluminometer (mode 535, Los Alamos Diagnostics) to which was attached a circulating water bath (model 2209, LKB) that held the tube at 25°C or 36°C. Control and all chemical test assays were performed by the addition of $CH_3SeSeCH_3$, CH_3SeOOH, or CH_3SSCH_3 directly to the luminometer test tube in the counting chamber. Superoxide dismutase (SOD; Sigma Chemical), used to quench the CL reactions, was added to the test tube before the addition of the selenium or sulfur compound. WARNING: Dimethyldiselenide and dimethyldisulfide are labeled highly toxic, and dimethyldiselenide is particularly noxious. The toxicity of methylseleninic acid has not been fully investigated.

Results

Under the conditions of the experiments described above, the complete CL cocktail produced very low, yet detectable, amounts of background CL at pH 9.2 and 7.4. This is due to the ambient spontaneous oxidation of GSH and was not significant to the tests performed. At both pH levels, the sele-

nium and sulfur compounds, in the presence of lucigenin alone (without GSH), produced no additional CL. Similarly, the selenium and sulfur compounds, in the presence of GSH alone (without lucigenin), produced no CL. We previously showed that native SOD quenches the CL generated, whereas heated denatured SOD does not quench the CL. Thus the CL produced and quantitated in these assays can be attributed to the generation of O_2^-. We also previously demonstrated that the CL generated by selenoate catalysis may be used to quantitatively approximate the amount of catalytic selenium present (24).

As shown in Fig. 2, $CH_3SeSeCH_3$ generated O_2^- at pH 9.2. The ability to measure 1.67 mg of $CH_3SeSeCH_3$ activity was preceded by a burst of high-intensity CL during the first 30 s, which could not be quantified. The addition of 100 U of SOD before the addition of $CH_3SeSeCH_3$ quenched the burst of CL activity seen with $CH_3SeSeCH_3$ alone. Figure 3 shows that, under the same conditions, $CH_3SeSeCH_3$ produced CL at a more physiological pH 7.4, albeit with less CL activity and without the initial CL burst. One hundred units of SOD quenched most of the CL activity at pH 7.4, as well as at pH 9.2, the optimum catalytic pH for the SOD enzyme.

Figures 4 and 5 show experiments identical to those described above, conducted using CH_3SSCH_3 at pH 9.2 and 7.4, respectively. On an absolute CL basis, only a fractional amount of CL was generated by CH_3SSCH_3 compared with $CH_3SeSeCH_3$ at pH 9.2 and 7.4. Additionally, SOD had no effect on the small amount of CL that was generated by CH_3SSCH_3 at the pH levels tested. Figures 6 and 7 display the CL activities of 1.0 μmol of $CH_3SeSeCH_3$ and 1.0 μmol of CH_3SSCH_3 at pH 9.2 and 7.4, respectively, with and without the addition of SOD. At these concentrations, only

Generation of Superoxide by Dimethyldiselenide at pH 9.2

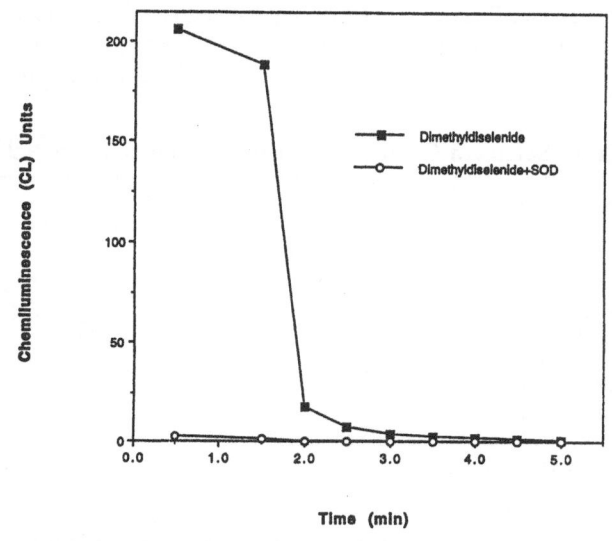

Figure 2. Generation of superoxide by dimethyldiselenide at pH 9.2 as measured by lucigenin-dependent chemiluminescence (CL). CL produced by 1.67 mg of dimethyldiselenide in the presence and absence of 100 U of superoxide dismutase (SOD) was measured in 1.0 ml of sodium borate buffer containing 4.0 mg of reduced GSH and 20 μg of lucigenin.

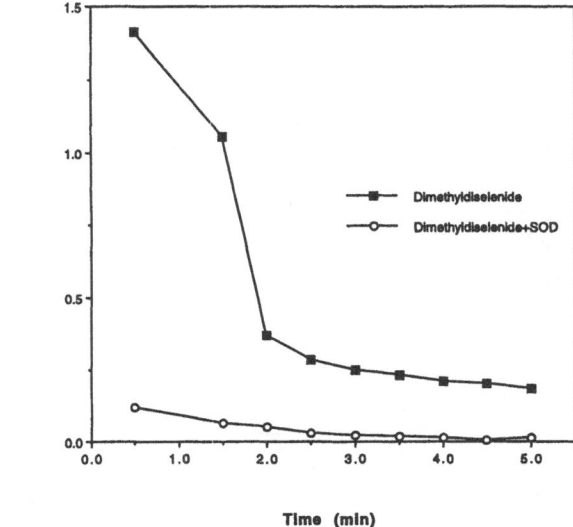

Figure 3. Generation of superoxide by dimethyldiselenide at pH 7.4 as measured by lucigenin-dependent CL. CL produced by 1.67 mg of dimethyldiselenide in the presence and absence of 100 U of SOD was measured in 1.0 ml of sodium borate buffer containing 4.0 mg of reduced GSH and 20 μg of lucigenin at 25°C.

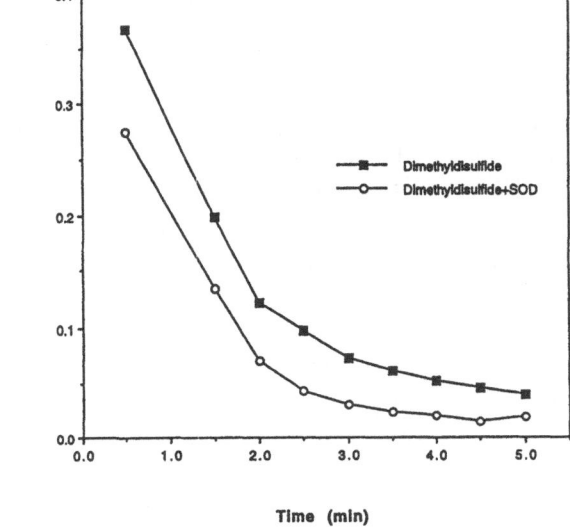

Figure 4. Generation of superoxide by dimethyldisulfide at pH 9.2 as measured by lucigenin-dependent CL. CL produced by 1.67 mg of dimethyldisulfide in the presence and absence of 100 U of SOD was measured in 1.0 ml of sodium borate buffer containing 4.0 mg of reduced GSH and 20 μg of lucigenin at 25°C.

$CH_3SeSeCH_3$ generated measurable CL at either pH. Production of CL by 1.0 μmol (0.188 mg) of $CH_3SeSeCH_3$ at pH 9.2 (Fig. 5), the optimal catalytic pH, was approximately three times higher than the CL generated by 17.8 μmol (1.67 mg) of CH_3SSCH_3 at the same pH (Fig. 4). CH_3SeOOH provided to us in distilled water at a selenium concentration of 22.7 mM was assayed under conditions identical to those de-

scribed above, except the temperature of the cuvette was raised to 36°C. Dilutions of CH_3SeOOH in distilled water were assayed for the ability to generate $O_2^-\cdot$. Figure 8 shows the activity of CH_3SeOOH-generated CL at 2.27 μmol of selenium with and without added SOD. As with $CH_3SeSeCH_3$, SOD completely quenched the CL generated. Figure 9 shows that CL from the addition of CH_3SeOOH could be de-

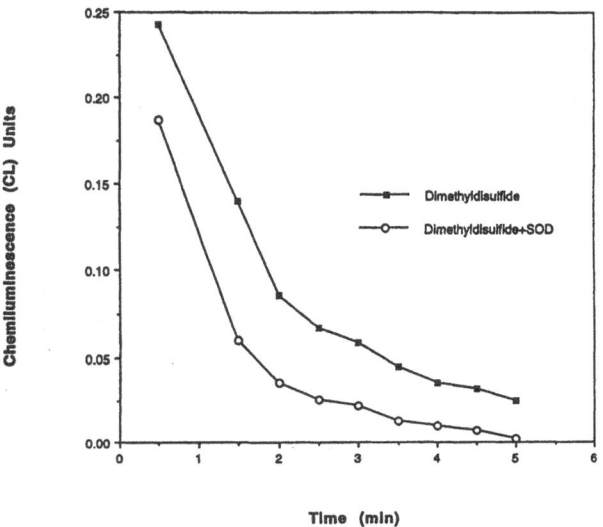

Figure 5. Generation of superoxide by dimethyldisulfide at pH 7.4 as measured by lucigenin-dependent CL. CL produced by 1.67 mg of dimethyldisulfide in the presence and absence of 100 U of SOD was measured in 1.0 ml of sodium borate buffer containing 4.0 mg of reduced GSH and 20 μg of lucigenin at 25°C.

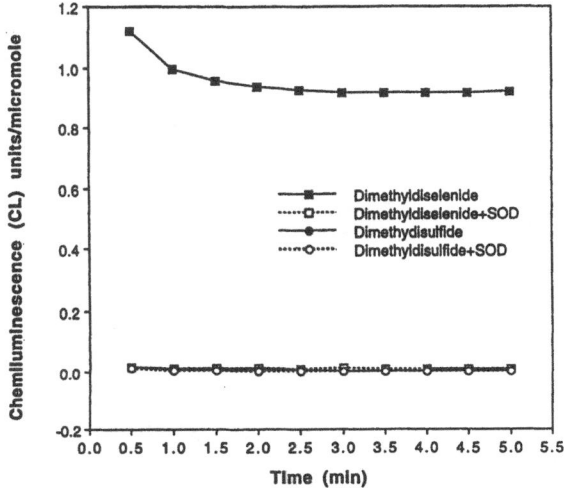

Figure 6. Generation of superoxide by equimolar amounts of dimethyldiselenide and dimethyldisulfide at pH 9.2 as measured by lucigenin-dependent CL. CL produced by 1.0 μmol of dimethyldiselenide and dimethyldisulfide in the presence and absence of 100 U of SOD was measured in 1.0 ml of sodium borate buffer containing 4.0 mg of reduced GSH and 20 μg of lucigenin at 25°C.

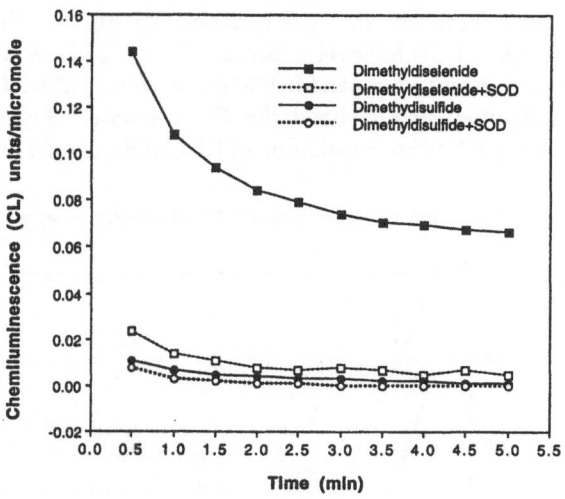

Generation of Superoxide at pH 7.4

Figure 7. Generation of superoxide by equimolar amounts of dimethyl-diselenide and dimethyldisulfide at pH 7.4 as measured by lucigenin-dependent CL. CL produced by 1.0 μmol of dimethyldiselenide and dimethyldisulfide in the presence and absence of 100 U of SOD was measured in 1.0 ml of sodium borate buffer containing 4.0 mg of reduced GSH and 20 μg of lucigenin at 25°C.

tected from background CL down to a level of 0.56 nmol of selenium.

Discussion

Selenium compounds are known to be toxic to cells in vitro as well as in vivo, and absolute toxicity depends on the chemical form of selenium, its concentration, and its metabolism (25,26). In general, selenite and diselenides are very toxic to cells in culture and to animals in vivo, whereas L-se-lenomethionine and L-Se-methylselenocysteine are less toxic to cells in culture (26), and these selenium compounds are not very toxic in vivo relative to selenite or diselenides (26,27). In cell culture, selenite induces DNA laddering and apoptosis in cells (28–31), whereas much higher concentrations of L-selenomethionine and L-Se-methylselenocysteine, 8- to 10-fold or more, are required to induce toxicity and apoptosis (31).

In vitro, selenite and all diselenides that we have examined and tested under the experimental conditions described here generate $O_2^-\cdot$, as measured by CL (Table 1), and the CL is quenched by the addition of SOD (19). The ability of diselenides to be reduced by GSH, as well as other thiols, forming RSe^-, has been elucidated in detail by Chaudiere et al. (32) (Fig. 10). RSe^- is seen to redox cycle indefinitely in the presence of GSH and oxygen, continuously generating $O_2^-\cdot$ and H_2O_2. It is this redox cycling of RSe^- that appears to account for selenium's toxicity in vitro as well as in vivo.

Unlike diselenides, the monoselenide dietary amino acids L-selenomethionine ($CH_3SeCH_2CH_2CHNH_2COOH$) and L-Se-methylselenocysteine ($CH_3SeCH_2CHNH_2COOH$) are not reducible to RSe^- by GSH or other thiols such as di-thiothreitol in vitro, and therefore they do not redox cycle in vitro (Table 1). These selenoethers do not generate $O_2^-\cdot$ in the in vitro CL assay and, as noted above, are much less toxic to cells in vitro and to animals in vivo. L-Seleno-methionine and L-Se-methylselenocysteine are known, however, to be toxic to cells in vitro at much higher selenium concentrations than is selenite and in vivo to animals at much higher than normal selenium dietary levels (33). As shown by the many experiments of Ip et al. (14,15) and as reviewed by Ip (16), the monomethyl species of selenium (CH_3SeH) must be continuously generated from selenium compounds to have carcinostatic activity. This conclusion

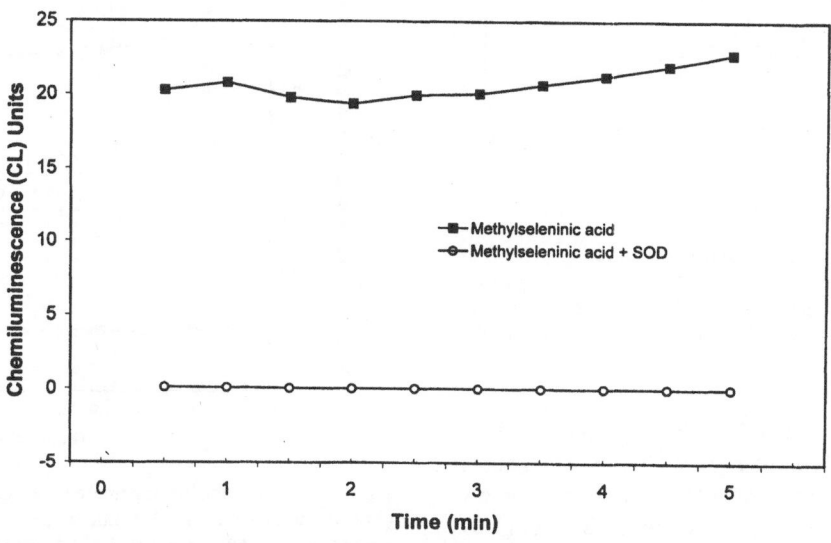

Generation of Superoxide by Methylseleninic Acid at pH 9.2

Figure 8. Generation of superoxide by methylseleninic acid at pH 9.2 as measured by lucigenin-dependent CL. CL produced by 0.227 μmol of selenium in the form of methylseleninic acid in the presence and absence of 100 U of SOD was measured in 1.0 ml of sodium borate buffer containing 4.0 mg of reduced GSH and 20 μg of lucigenin at 36°C.

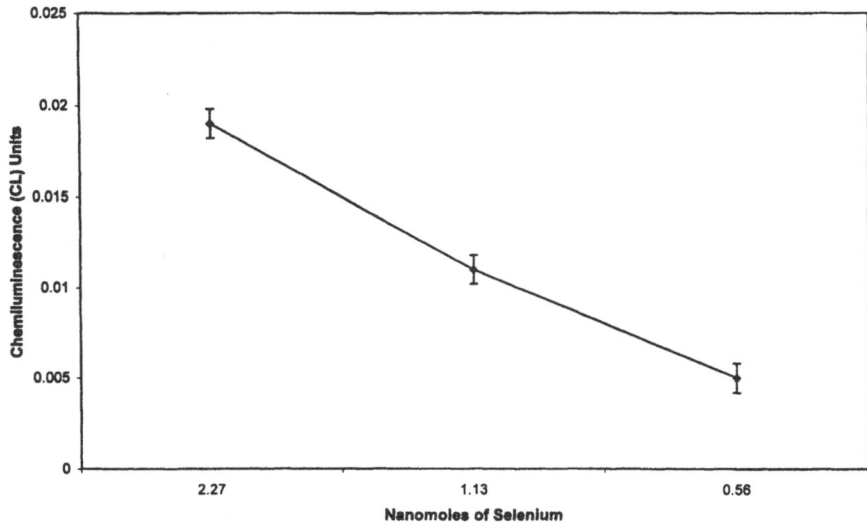

Figure 9. Generation of superoxide by nanomolar amounts of selenium in the form of methylseleninic acid at pH 9.2 as measured by lucigenin-dependent CL. CL produced by 2.27, 1.13, and 0.56 nmol of selenium in the form of methylseleninic acid was measured in 1.0 ml of sodium borate buffer containing 4.0 mg of reduced GSH and 20 μg of lucigenin at 36°C. Values are means ± SEM of repeated 30-s measurements ($n = 17$). Mean CL values for all 3 levels of selenium were significantly different from those produced by a reagent blank devoid of added selenium ($P < 0.01$).

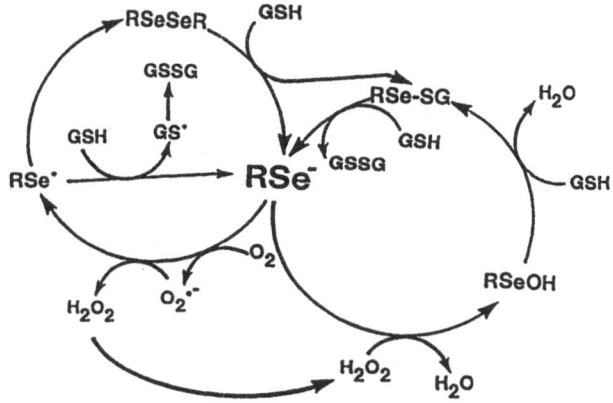

Figure 10. Free radical redox cycle of selenoate anions. [From Chaudiere et al. (32).]

was derived from the testing of the carcinostatic activity of a number of selenium metabolites reported by Ip (16) in the rat mammary cancer model. The best organic selenium compound for the prevention of mammary cancer was L-Se-methylselenocysteine, followed by dimethylselenide and then selenobetaine. The least effective metabolite was the trimethylselenonium ion, the major excretory product of selenium metabolism in animals and humans.

Our experience with redox-cycling, $O_2^-\cdot$-generating diselenide selenium compounds in the presence of GSH suggested to us that the monomethylated selenium species CH_3SeH, formed on the reduction of $CH_3SeSeCH_3$ and CH_3SeOOH by GSH, should be a highly reactive redox-cycling species in its ionized form, CH_3Se^- (the pK_a of the selenol of selenocysteine being 5.25). As described in **Results,** we found CH_3SeH, derived from the reduction of $CH_3SeSeCH_3$ or CH_3SeOOH, to be a very active redox com-

pound producing $O_2^-\cdot$. Our experience also suggests that this selenonium species, CH_3SeH, is the most catalytically active of any diselenide reduced by GSH that we have tested. In contrast, the corresponding sulfur species, methylthiol (CH_3S^-, the pK_a of the thiol of cysteine being ~8–9) generated only fractional amounts of CL in comparison to CH_3SeH. The very small amount of CL generated by CH_3S^- appeared not to be due to $O_2^-\cdot$, inasmuch as SOD did not significantly quench the very small amount of CL generated. This CL may be due to the transient generation of some other free radical species on reduction with GSH, such as a thiol radical, i.e., $CH_3S\cdot$ or $GS\cdot$. The inability of CH_3S^- as well as other thiols, i.e., cysteine, that we have tested to redox cycle may explain why these sulfur compounds are not nearly as toxic or as effective as selenols in the prevention of cancer (34).

The results of the experiments of Ip et al. and other in vitro and in vivo selenium experiments provide a plausible understanding of how the dietary selenium compounds L-selenomethionine and L-Se-methylselenocysteine are metabolized to produce an RSe^- selenium species, which can redox cycle to induce apoptosis (27–31,35) and cancer cell death by the initial generation of $O_2^-\cdot$. It is metabolically possible that two redox selenium species (RSe^- and CH_3Se^-) of L-selenomethionine and L-Se-methylselenocysteine could be formed in vivo. The amino acid RSe^- could be formed by a demethylation reaction and would be fully expected to redox cycle, generating $O_2^-\cdot$. In fact, demethylation in vivo of methylated selenium compounds, as shown by Ip, is known to occur (11).

Monomethylated and dimethylated selenium compounds, and even the trimethylselenonium ion, if present in the diet in sufficient concentration, can support the synthesis of the

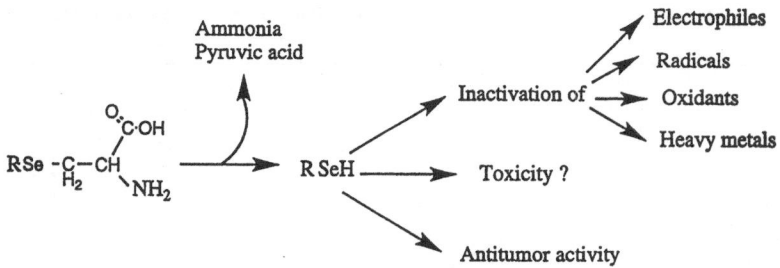

Figure 11. Schematic representation of β-lyase-catalyzed β-elimination reaction of selenium-substituted selenocysteine conjugates forming selenol compounds with potential chemoprotective, antitumor, or toxic properties. [Reprinted with permission from I Andreadou, BM Wiro, JNM Commandeur, EA Worthington, and NPE Vermeulen: Synthesis of novel Se-substituted selenocysteine derivatives as potential kidney selective prodrugs of biologically active selenol compounds: evaluation of kinetics of β-elimination reactions in rat renal cytosol. *J Med Chem* **39**, 2040–2046. Copyright (1999) American Chemical Society.]

selenoenzyme glutathione peroxidase (14). Glutathione peroxidase can be synthesized only if selenium demethylation reactions take place to form H_2Se (Fig. 1), as has been demonstrated for the trimethylselenonium ion. This selenonium anion of the selenoamino acids is not as likely to be formed in cancer cells because of the reported presence of high levels of β-lyases (36). More likely to be formed is CH_3SeH, as shown by Ip et al. in their studies of the selenium-containing metabolites in the prevention of mammary carcinogenesis (16).

This metabolic formation of CH_3SeH is also suggested by Andreadou et al. (36) in their studies of nontoxic Se- (alkyl)-Se conjugates, i.e., prodrugs that contain selenoethers (RSeR) directed to renal carcinoma (Fig. 11). Highly active β-lyases are reported to be prevalent in renal carcinoma cells, and perhaps they are also present in other cancer cells, forming the basis for the generation of CH_3SeH de novo from L-selenomethionine and L-Se-methylselenocysteine. This event would likely be similar to the 14 selenocysteine-substituted conjugates studied by Andreadou et al. in renal carcinoma cells, where β-lyase activity and carcinostatic activity were inhibited by the addition of the β-lyase inhibitor aminooxyacetic acid. It seems highly likely, therefore, that the dietary selenoether species CH_3SeR, L-selenomethionine, and L-Se-methylselenocysteine need only to be delivered to cancer cells continuously at sufficient concentrations to generate the highly catalytic CH_3SeH in the presence of β-lyase activity to produce $O_2^-\cdot$, inducing oxidative stress and apoptosis and, over time, having carcinostatic activity.

Ip and co-workers (37) have indicated that L-Se-methylselenocysteine is more effective as a chemopreventative selenium nutraceutical than L-selenomethionine in their DMBA animal model of mammary carcinogenesis. Several factors likely account for these observations. The first is that L-selenomethionine is incorporated into the primary protein structure in the place of methionine, effectively reducing its availability to generate CH_3SeH, whereas L-Se-methylselenocysteine is a non-protein-accumulating selenoamino acid and is fully available to generate CH_3SeH. There may also be differences in these two selenoamino acids as substrates for β-lyase cleavage. Additionally, there may also be an alternative pathway of generating CH_3SeH, as suggested by Martin (38) in 1979. Martin suggested that L-Se-

methylselenocysteine might follow the metabolic pathway of S-methylcysteine, the hydrolysis of which would yield serine and CH_3SeH. Thus there are several metabolic possibilities for the generation of CH_3SeH from L-selenomethionine and L-Se-methylselenocysteine. The metabolism of these two selenoamino acids needs to be examined in detail in vivo for CH_3SeH formation, induction of oxidative stress, and the mechanism by which they invoke carcinostasis.

One final experimental question concerning the association of $O_2^-\cdot$ generation and toxicity CH_3SeH ex vivo and in vivo needed to be answered experimentally in vitro: Could $O_2^-\cdot$ be detected in vitro from the catalytic activity of CH_3SeH at selenium levels found to be toxic to cells ex vivo? The answer is a definitive yes. As shown in Fig. 8, we were able to measure $O_2^-\cdot$ generation by CH_3SeOOH in the presence of GSH at selenium levels that induced apoptosis in cultured cells. Thus there is no experimental contradiction or doubt that CH_3SeH can generate detectable amounts of $O_2^-\cdot$ at levels that may induce oxidative stress in cells and ultimately produce apoptosis and cell death. There is one additional line of evidence that suggests that this is true. We have covalently attached a selenium molecule of the configuration RSe^- to a variety of antibodies, peptides, a steroid, and even polymeric surfaces, and all were shown to generate $O_2^-\cdot$ in vitro as described here (39). In addition, these site-directed or surface selenols are toxic to cells only if bound to them, in very close proximity, or are phagocytized. This suggests to us experimentally that there exists a close, if not identical, cause of cell death from dietary seleno amino acids used for chemoprevention and selenium-bound vector molecules that may be used for chemotherapy.

Acknowledgments and Notes

This work is dedicated to a friend and colleague, Larry Clark, who succumbed to a disease he worked so hard to try to prevent. The work described here has been published in abstract form (33). Address correspondence to Julian Spallholz, Ph.D., Food and Nutrition, College of Human Sciences, Texas Tech University, Lubbock, TX 79413. E-mail: jspallholz@hs.ttu.edu.

The following references published since submission of this manuscript are relevant to the data presented herein: Jung U, Zheng X, Yoon S, and Chung A: Se-methylselenocysteine induces apoptosis mediated by reactive oxygen species in HL-60 cells. *Free Radic Biol Med* **31**, 479–489, 2001; Ip C and Dong Y: Methylselenocysteine modulates proliferation and apoptosis biomarkers in premalignant lesions of the rat mammary gland. *Anticancer Res* **21**, 863–867, 2001; Kim T, Jung U, Cho DY, and Chung AS: Se-methylselenocysteine induces apoptosis through caspase activation in HL-60 cells. *Carcinogenesis* **22**, 559–565, 2001; and Unni E, Singh U, Ganther HE, and Sinha R: Se-methylselenocysteine activates caspase-3 in mouse mammary epithelial tumor cells in vitro. *Biofactors* **14**, 169–177, 2001. The following reference should have been cited: Ip C, Thompson HJ, Zhu Z, and Ganther HE: In vitro and in vivo studies of methylseleninic acid: evidence that a monomethylated selenium metabolite is critical for cancer chemoprevention. *Cancer Res* **60**, 2882–2886, 2000.

Submitted 31 October 2000; accepted in final form 20 November 2000.

References

1. Burk RF: *Selenium in Biology and Human Health.* New York: Springer-Verlag, 1994.

2. Bock A and Forchammer J: Selenocysteine: the 21st amino acid. *Mol Microbiol* **5**, 515–520, 1991.

3. Wen H, Shi B, Boylan LM, Chen JJ, Davis R, et al.: A comparison of the bioavailability of selenium from meats, poultry, seafoods fed at adequate (0.10 ppm) and inadequate (0.05 ppm) dietary levels to selenium-deficient Fischer 344 rats. *Biol Trace Elem Res* **58**, 45–58, 1997.

4. *Recommended Dietary Allowances*, 10th ed. Washington, DC: National Academy Press, 1989.

5. Clark LC, Combs GF, Turnbull BW, Slate EH, Chalker DK, et al.: Effects of selenium supplementation for cancer prevention in patients with carcinoma of the skin. *JAMA* **276**, 1957–1963, 1996.

6. Watrach AM, Milner JA, and Watrach MA: Effect of selenium on growth rate of canine mammary carcinoma cells in athymic nude mice. *Cancer Lett* **15**, 137–149, 1982.

7. Poirier KA and Milner JA: Factors influencing the antitumorigenic properties of selenium in mice. *J Nutr* **113**, 2147–2154, 1983.

8. Milner JA, Greeder GA, and Poirier KA: Selenium and transplantable tumors. In *Selenium in Biology and Medicine*, Spallholz JE, Martin JL, and Ganther HE (eds). Westport, CT: AVI, 1981.

9. Schrauzer GN: Selenium: mechanistic aspects of anticarcinogen action. *Biol Trace Elem Res* **33**, 51–62, 1992.

10. Yan L, Yee JA, McGuire MH, and Graef GL: Effect of dietary supplementation of selenite on pulmonary metastasis of melanoma cells in mice. *Nutr Cancer* **28**, 165–169, 1996.

11. Ganther HE and Lawrence JR: Chemical transformations of selenium in living organisms. Improved forms of selenium for cancer prevention. *Tetrahedron* **53**, 12229–12310, 1997.

12. Lee BJ, Rajagonpalan M, Kim YS, You KH, Jacobson KB, et al.: Selenocysteine tRNA [Ser] *sec* gene is ubiquitous within the animal kingdom. *Mol Cell Biol* **10**, 1940–1949, 1990.

13. Foster SJ, Kraus RJ, and Ganther HE: The metabolism of selenomethionine, Se-methylselenocysteine, their selenonium derivatives, and trimethylselenonium in the rat. *Arch Biochem Biophys* **251**, 77–86, 1986.

14. Ip C, Hayes C, Budnick RM, and Ganther HE: Chemical form of selenium, critical metabolites, and cancer prevention. *Cancer Res* **51**, 595–600, 1991.

15. Vadhanavikit S, Ip C, and Ganther HE: Metabolites of sodium selenite and methylated selenium compounds administered at cancer chemoprevention levels in the rat. *Xenobiotica* **23**, 731–745, 1993.

16. Ip C: Lessons from basic research in selenium and cancer prevention. *J Nutr* **128**, 1845–1854, 1998.

17. Ganther HE: Selenium metabolism, selenoproteins and mechanisms of cancer prevention: complexities with thioredoxin reductase. *Carcinogenesis* **20**, 1657–1666, 1999.

18. Jiang C, Jiang W, Ip C, Ganther H, and Lu J: Selenium-induced inhibition of angiogenesis in mammary cancer at chemopreventive levels of intake. *Mol Carcinogenesis* **26**, 213–225, 1999.

19. Spallholz JE: On the nature of selenium toxicity and carcinostatic activity. *Free Radic Biol Med* **17**, 45–64, 1994.

20. Spallholz JE: Free radical generation by selenium compounds and their prooxidant toxicity. *Biomed Environ Sci* **10**, 260–270, 1997.

21. Yan L and Spallholz JE: Generation of reactive oxygen species from the reaction of selenium compounds with thiols and mammary tumor cells. *Biochem Pharmacol* **45**, 429–437, 1993.

22. Xu H, Feng Z, and Yi C: Free radical mechanism of the toxicity of selenium compounds. *Huzahong Longong Daxue Xuebao* **19**, 13–19, 1991.

23. Faulkner K and Fridovich I: Luminol and lucigenin as detectors and O_2^-. *Free Radic Biol Med* **15**, 447–451, 1993.

24. Chen L and Spallholz JE: Preparation of a cytolytic selenium antibody immunoconjugate. *Free Radic Biol Med* **19**, 713–724, 1994.

25. Wilson AC, Thompson HJ, Schedin PJ, Gibson NW, and Ganther HE: Effect of methylated forms of selenium on cell viability and the induction of DNA strand breakage. *Pharmacology* **43**, 1113–1141, 1992.

26. Stewart MS, Spallholz JE, Neldner KH, and Pence BC: Selenium compounds have disparate abilities to impose oxidative stress and induce apoptosis. *Free Radic Biol Med* **26**, 42–48, 1999.

27. Lanfear J, Fleming J, Wu L, Webster G, and Harrison PR: The selenium metabolite selenodiglutathione induces p53 and apoptosis: relevance to the chemopreventative effects of selenium? *Carcinogenesis* **15**, 1387–1392, 1994.

28. Garberg P, Stahl A, Worholm M, and Holberg J: Studies on the role of DNA fragmentation in selenium toxicity. *Biochem Pharmacol* **37**, 3401–3406, 1998.

29. Lu J, Kaek M, Jiang C, Webster G, and Harrison PR: Selenite induction of DNA strand breaks and apoptosis in mouse leukemic L1210 cells. *Biochem Pharmacol* **47**, 1531–1535, 1994.

30. Stewart MS, Davis RL, Walsh LP, and Pence BC: Induction of differentiation and apoptosis by sodium selenite in human colonic carcinoma cells (HT29). *Cancer Lett* **117**, 35–40, 1997.

31. Stewart MS, Spallholz JE, Neldner KH, and Pence B: Selenium compounds have disparate abilities to impose oxidative stress and induce apoptosis. *Free Radic Biol Med* **26**, 42–48, 1999.

32. Chaudiere J, Courtin O, and Leclaire J: Glutathione oxidase activity of selenocystamine: a mechanistic study. *Arch Biochem Biophys* **296**, 328–336, 1992.

33. Spallholz JE, Shriver BJ, and Reid TW: Dimethyldiselenide generates superoxide in an in vitro assay in the presence of glutathione (abstr). *FASEB J* **12**, A470, 1998.

34. Ip C and Ganther HE: Comparison of selenium and sulfur analogs in cancer prevention. *Carcinogenesis* **13**, 1167–1170, 1992.

35. Redman C, Xu MJ, Peng YM, Scott JA, Payne C, et al.: Involvement of polyamines in selenomethionine-induced apoptosis and mitotic alterations in human tumor cells. *Carcinogenesis* **18**, 1195–1202, 1997.

36. Andreadou I, Wiro BM, Commandeur JNM, Worthington EA, and Vermeulen NPE: Synthesis of novel Se-substituted selenocysteine derivatives as potential kidney selective prodrugs of biologically active selenol compounds: evaluation of kinetics of β-elimination reactions in rat renal cytosol. *J Med Chem* **39**, 2040–2046, 1996.

37. Ip C, Birringer M, Block E, Kotrebai M, Tyson TF, et al.: Chemical speciation influences comparative activity of selenium-enriched garlic and yeast in mammary cancer prevention. *J Agric Food Chem* **48**, 2062–2070, 2000.

38. Martin J: Assimilation and metabolism of organoselenium compounds. *Proceeding of the Symposium, Metz, France, 1979*, pp 133–144.

39. Spallholz JE, Boylan LM, Shriver BJ, and Reid TW: Selenium free radical chemistry: applications for pharmaceuticals. *Proceedings of the 6th Annual Selenium Telurium Development Association Symposium on the Uses of Selenium and Tellurium, 1998*, pp 47–56.

NUTRITION AND CANCER, *40*(1), 42–49

Molecular Mechanisms of Cancer Prevention by Selenium Compounds

Janis Fleming, Aurnab Ghose, and Paul R. Harrison

Abstract: Selenium compounds that are chemopreventive in animal models inhibit cell growth and induce apoptosis in vitro, and this could explain how they reduce the outgrowth of tumor cells in vivo. Our recent work has shown that primary cultures of oral carcinoma biopsies are significantly more sensitive than normal oral mucosa cultures to induction of apoptosis by a natural selenium metabolite [seleno-diglutathione (SDG)], and this is associated with induction of Fas ligand, a well-known mediator of apoptosis in other contexts, and activation of so-called stress kinase signaling pathways, particularly the Jun NH$_2$-terminal kinase (JNK). Heme oxygenase, another marker of stress responses, is also induced by selenite and SDG. The selective activation of the Fas pathway in carcinomas could be responsible directly for their destruction by apoptosis or target them for attack by immunologic responses. In contrast, although the potent pharmacological selenium chemopreventive agent 1,4-phenylenebis(methylene)selenocyanate (p-XSC) also induces Fas ligand, heme oxygenase, and stress kinase pathways, apoptosis/Fas induction is not so strongly JNK-dependent and p-XSC does not show tumor selectivity. These differences in mechanism between SDG and p-XSC may be due to the manner in which they induce redox changes in the cells, since although the effects of SDG and p-XSC are prevented by antioxidants such as glutathione or N-acetylcysteine, hydroxyl radical scavengers such as mannitol or pyrrolidine dithiocarbamate only protect against the effects of p-XSC.

Animal Evidence for a Cancer-Protective Effect of Selenium

There is very convincing evidence that a high dietary level of naturally occurring selenium, usually in the form of sodium selenite, substantially reduces the incidence of a wide variety of animal cancers under conditions where animal growth and health are not affected (1–3). Most studies have observed the maximum cancer-protective effect at nontoxic levels considerably higher than normal nutritional levels, but there is also more limited evidence that sub-nutritional selenium deprivation increases cancer risk (re-

viewed in Refs. 4 and 5). In animal models, the most effective chemopreventive agents seem to be those that are metabolized directly to selenols (6–9), including aliphatic or benzyl selenocyanates (10–12), particularly 1,4-phenylenebis(methylene)selenocyanate (p-XSC) (13–15), and Se-methylselenocysteine, which may be the major nonvolatile form of selenium in plant foods (16). In particular, selenium derivatives such as p-XSC are effective in animal carcinogenesis models in which selenite is not so effective, for example, 4-(methylnitrosamino)-(3-pyridyl)-1-butanone-induced lung cancers (13).

Epidemiological Evidence for a Cancer-Protective Effect of Selenium

Human epidemiological evidence also indicates a statistically significant inverse relationship between selenium level and risk of cancer overall, particularly in men (17–28; for review see Ref. 5). In terms of individual cancer types, the evidence from the largest studies is strongest for upper respiratory tract, prostate, esophageal, and stomach cancers (27,29–32), but there is little evidence of any link with colon (23,27,28,32) or breast (34–36) cancer risk in such cohort studies, although the length of follow-up in some of these studies was quite short. Three intervention trials have also investigated whether selenium supplementation reduces the risk of cancer. An early randomized intervention trial in China, with ~1,000 subjects in each of the treatment and placebo arms, reported that dietary supplementation with selenium at 200 µg/day as selenized yeast for 2 yr reduced the incidence of liver cancer in a high-risk area by 45% (P = 0.05) (37). A joint Chinese/US randomized factorial design study in a nutritionally poor population in Linxian County, China, showed that dietary supplementation for 5 yr with a cocktail of three antioxidants (β-carotene, vitamin E, and selenium) at about twice their Recommended Dietary Allowance produced a statistically significant reduction in overall cancer rate of 13% and a 21% reduction in stomach cancer rate, whereas supplementation with other cocktails of vitamins or trace elements had no effect (38,39). It seems plausible that selenium was responsible for the effect of the

---

The authors are affiliated with the Beatson Institute for Cancer Research, CRC Beatson Laboratories, Bearsden, Glasgow G61 1BD, Scotland, UK.

antioxidant cocktail, since evidence from intervention trials suggests that neither β-carotene nor vitamin E reduces cancer risk generally (40–43). Indeed, although there may be a significant inverse relationship between prostate cancer and vitamin E intake (44,45), β-carotene supplementation was consistently associated with a higher risk of lung cancers in subjects who were current smokers (46–48). The case for selenium supplementation reducing cancer risk has received greatest support from the recent randomized placebo-controlled intervention trial by Dr. Larry Clark and his colleagues involving 1,312 men with a prior history of skin basal or squamous cell carcinoma (49,50). Although they found no evidence for a reduction in risk of recurrence of skin cancer with selenium at 200 µg/day as selenized yeast, total cancer rates in the treatment arm were reduced significantly (by 37%), as were the individual rates of lung, colorectal, and prostate cancers (49,50). The strongest treatment effect was observed in subjects with the lowest plasma selenium levels before supplementation.

Possible Molecular Mechanisms

The molecular mechanism(s) responsible for reduction in cancer risk by selenium is unclear. The early hypothesis that the chemopreventive effect of selenium might be mediated by selenoproteins, such as the glutathione peroxidases, eliminating tumor-promoting reactive oxygen species, now seems unlikely, since the activities of known selenoproteins seem to be saturated in animals at a much lower dietary selenium level (0.1–0.4 ppm) (51–53) than that required for the maximum chemopreventive effect of selenium (2–4 ppm) (1–4). There is further evidence that at least the cytosolic glutathione peroxidase (*Gpx1*) is not involved, since *Gpx1*-null mice do not show any abnormal histopathologies up to 15 mo of age (54,55) [although they are more sensitive to exogenous oxidative stress (56)], whereas *Gpx1*-transgenic mice are actually more sensitive to dimethylbenz[*a*]anthracene-12-*O*-tetradecanoylphorbol 12-acetate-induced skin cancer, rather than the reverse, hypothetically because the generation of tumor-promoting lipoxygenase-derived peroxides is increased (57). However, thioredoxin reductase remains a possible candidate, since it has been shown to be inhibited by continuous high selenium levels, possibly because of diselenide bond formation at the selenocysteine in the active site (58). New selenoproteins have recently been discovered (59,60), but their functions and selenium saturation levels are unknown.

Although selenite and p-XSC can reduce DNA damage by carcinogens (4,61–63), they are also effective if given after the carcinogen in the early phase of tumor progression (1,4), suggesting that they may act primarily as antipromotion agents. This is also suggested by the fact that, in general, the relative efficacies of selenium derivatives as chemopreventive agents in vivo parallel their growth-inhibitory effects in vitro and their ability to induce apoptosis (64–68). Moreover, induction of apoptosis by selenium metabolites in vitro is readily detectable at levels toward the up-

per limit of plasma concentrations found in humans, i.e., ~5 µM, although higher than the average level (Fig. 1). Thus, understanding the mechanisms mediating the growth-inhibitory effects in vitro may be relevant to the mechanism of chemoprevention in vivo, and so identifying the molecular target(s) of selenium action may suggest avenues for devising novel compounds that may be useful as cancer-protective agents, for example, in high-risk groups. We previously showed that treatment of cells with the selenite metabolite selenodiglutathione (SDG) induces p53 accumulation in cells containing wild-type p53 (65). However, as we and others have shown, selenium compounds induce growth arrest/apoptosis of cells expressing mutant p53 (65) or lacking p53 completely (69); thus the mechanism whereby selenium compounds inhibit cell growth and induce apoptosis does not necessarily require a functional wild-type p53 pathway. Recent work showing that inorganic and organic forms of selenium induce *gadd34*, *gadd45*, and *gadd153* (69) suggests that interference with cell cycle check point controls is associated with growth arrest/apoptosis induced by selenium compounds.

Induction of Stress Responses by Interference With Redox Control

The main purpose of our recent work has been to identify the signaling mechanisms responsible for apoptosis induced by cancer-protective selenium compounds. Our experimental rationale was based on previous work by us and others that selenium compounds with the strongest cancer-protective properties in vivo, such as p-XSC, and metabolites of selenite, such as SDG, are also the most effective inducers of growth arrest and apoptosis in vitro. The reason for using SDG is based on our observation that selenite freshly added to cells only inhibited their growth after a lag period of 24 h, whereas conditioned medium from such cells transferred to

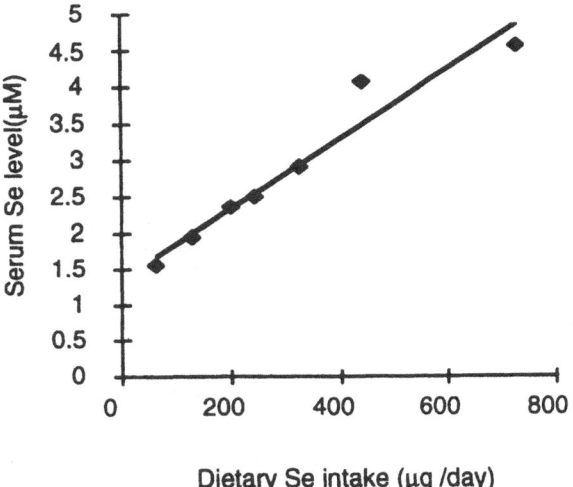

Figure 1. Range of selenium levels in vivo. [Data from Longnecker et al. (90).] Plotting selenium intake of people (µg/day) living in seleniferous areas against measured serum selenium levels (µM) revealed that serum concentrations of 3–5 µM selenium can be achieved, in vivo, by people taking high doses of selenium without any apparent adverse effects.

fresh cells to give the same selenium concentration inhibited their growth rapidly, and the active agent was shown to have a low molecular weight by passage through filters with different pore sizes (data not shown). The lag period was also reduced by the addition of glutathione together with selenite (Fig. 2). Both of these findings suggested that the active inhibitor might be SDG, which would be formed by reaction of selenite with glutathione intra- or extracellularly, and this was consistent with the fact that purified SDG inhibited cell growth without a significant lag period (Fig. 2). This growth arrest was also shown to be associated with apoptosis as determined by the TdT-mediated dUTP nick end labeling assay (data not shown).

Exactly how selenium compounds induce apoptosis is not clear. It has been postulated that selenium compounds induce oxidative stress by production of superoxide radicals or hydrogen peroxide (70–72), although our previous work suggested that SDG does not induce apoptosis in the same way as hydrogen peroxide (73), and this is supported by our new evidence that selenium compounds do not induce the same large-scale changes in phosphorylation of tyrosine residues in proteins characteristic of oxidants, such as hydrogen peroxide and diamide (Fig. 3). Nevertheless, selenite, SDG, and p-XSC induce heme oxygenase (Fig. 4), a well-established indicator of a variety of cellular stresses, for example, by oxidants, irradiation, and sulfhydryl reagents, as well as heme (74), and this occurs at selenite and SDG concentrations of 3–5 μM, which is within the range of plasma concentrations found in humans (Fig. 1). We also present further evidence here that although the growth-inhibitory effects of SDG and p-XSC in animal cell lines are prevented by the antioxidant N-acetylcysteine, putative hydroxyl radical scavengers, such as mannitol and pyrrolidine dithiocarbamate, are more effective against p-XSC than SDG (Fig. 5). This suggests that SDG and p-XSC do not act by precisely the same mechanism. One hypothesis to explain induction of growth arrest/apoptosis by SDG is that it may alter the redox status of the cells by manipulating the level of a cellular reducing agent, such as thioredoxin, that has been implicated in growth control in various contexts and is overexpressed in many tumors (reviewed in Ref. 75).

Tumor-Selective Induction of Apoptosis by the Selenium Metabolite SDG

In view of the fact that lung and esophageal cancers are two of the cancer types in which the epidemiological evidence for a cancer-protective effect of selenium is strongest, in our recent work (76) we focused on human oral cancers, since they share risk factors similar to those of the lung and other head and neck cancers and represent one of the few experimental models where biopsies of normal tissue and lesions at various stages of cancer progression can be obtained and studied in primary culture (77,78). One of the novel findings to emerge from this work (76) is that human oral carcinomas are more sensitive to induction of apoptosis by SDG than normal oral mucosa cells. This is clearly of considerable relevance to understanding the cancer-protective effect of selenium compounds. p-XSC does not show this tumor selectivity; the reasons for this difference between SDG and p-XSC are unclear but appear to be connected with the relative importance of the various kinase signaling pathways that mediate the effects of the two selenium compounds (see below). Another recent report has also shown that the LNCaP prostate carcinoma cell line is more sensitive to growth inhibition and induction of apoptosis by selenite or selenomethionine than primary prostate cells, although the molecular signaling mechanisms responsible were not explored (79).

Involvement of the Fas and Stress Kinase Pathways in Induction of Apoptosis by Selenium Compounds

Another important finding we have made is that induction of apoptosis by SDG and p-XSC is associated with a large induction in Fas ligand (Fas-L) expression above the low basal level in untreated cells (76); Fas-L is produced in the soluble form, which could therefore be secreted and act

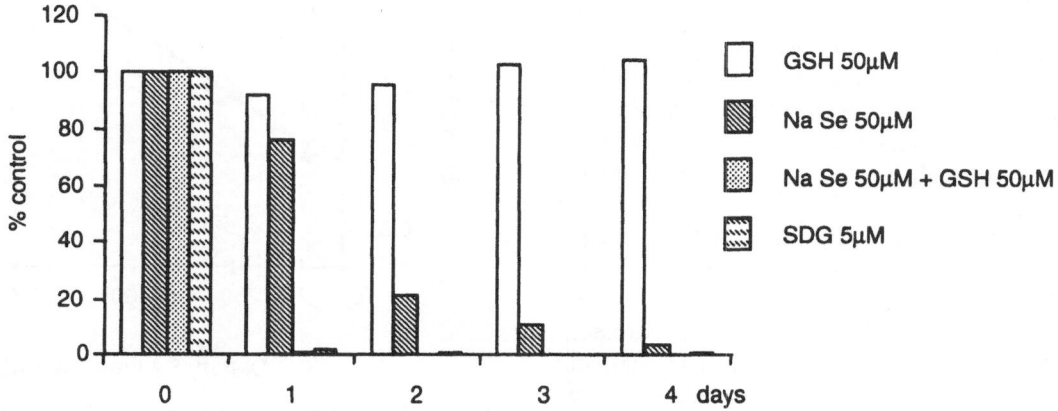

Figure 2. Potentiation of growth-inhibitory effects of selenite by glutathione (GSH): comparison with selenodiglutathione (SDG). Growth of C57 cells, a mouse mammary epithelial cell line, in response to treatment with 50 μM sodium selenite (Na Se), 50 μM GSH, 50 μM Na Se + 50 μM GSH, and 5 μM SDG for 1–4 days is shown. Cell viability was assayed using the 3-(4,5-dimethylthiazol-2-yl)-2,5-diphenyltetrazolium bromide (MTT) assay (Promega, UK) according to the manufacturer's protocol.

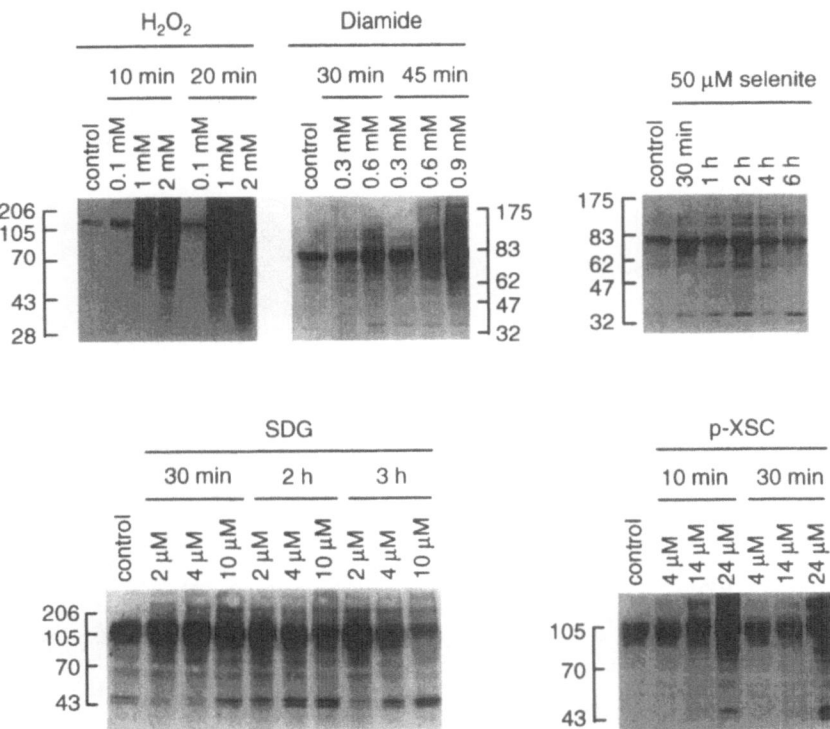

Figure 3. Tyrosine phosphorylation patterns in response to treatment with oxidants and selenium compounds. Cells were treated as indicated with H_2O_2, diamide, selenite, SDG, or 1,4-phenylenebis(methylene)selenocyanate (p-XSC), protein extracts were prepared, and immunoblot analysis was performed using a phosphotyrosine-specific antibody (Santa Cruz Biotechnology, Santa Cruz, CA).

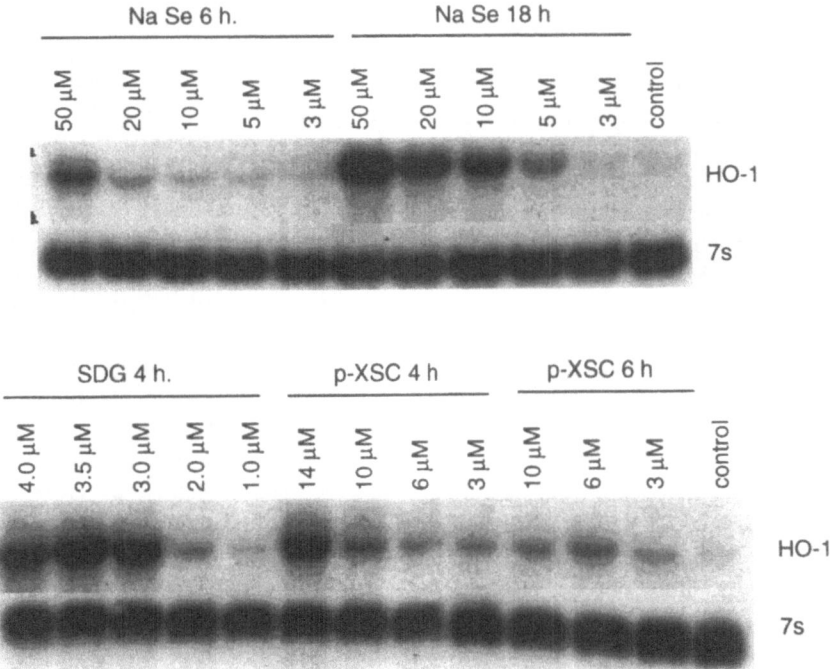

Figure 4. Induction of heme oxygenase (HO) mRNA by selenium compounds. C57 cells were treated with indicated concentrations of sodium selenite, SDG, or p-XSC for 4, 6, or 18 h, and RNA was extracted and analyzed by Northern transfer and hybridization with an HO-1 cDNA probe (74) or a 7S ribosomal RNA probe as a loading control.

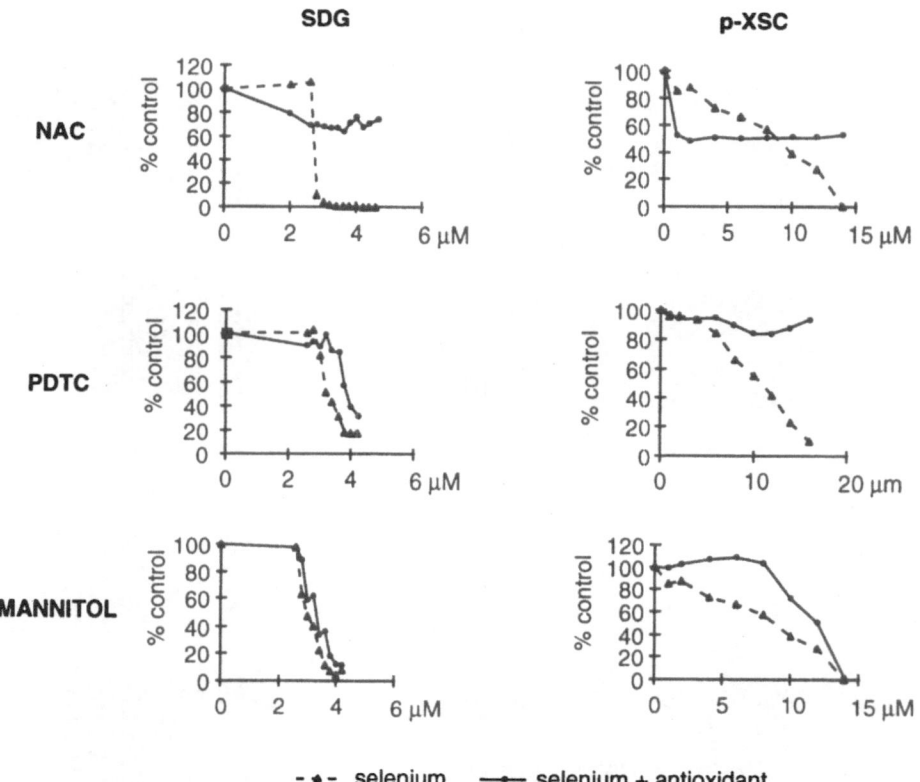

- ▲ - selenium   ——•—— selenium + antioxidant

Figure 5. Effects of antioxidants on inhibition of growth by selenium compounds. Cells were pretreated for 18 h with 20 mM *N*-acetyl cysteine (NAC) or 100 mM mannitol or for 2 h with 80 μM pyrrolidine dithiocarbamate (PDTC) and then with indicated concentrations of SDG or p-XSC in the continued presence of antioxidant for 4 h (NAC or PDTC) or 18 h (mannitol). Cells were then washed in phosphate-buffered saline, and cell growth was measured using the 3-(4,5-dimethylthiazol-2-yl)-2,5-diphenyltetrazolium bromide assay. Results are expressed relative to cells treated with antioxidant alone.

intercellularly. The extent of Fas-L induction by the selenium compounds correlates closely with the level of apoptosis induced in normal mucosa or carcinomas. Fas-L expression in biopsies of squamous carcinomas of the head and neck has also been demonstrated recently in another report and shown to be biologically active in inducing apoptosis in cocultivated activated T lymphocytes (80). Because normal oral mucosa and oral carcinomas express the Fas receptor constitutively (data not shown), induction of Fas-L may explain why selenium compounds induce apoptosis in human oral cells. In vivo, the soluble Fas-L produced by carcinoma cells could also enhance immunologic responses that could target the carcinoma cells (81). Activation of the Fas pathway by selenium compounds could therefore be a factor explaining their effects in enhancing antitumor immune responsiveness (5,82). Our data show that Fas-L induction by the selenium metabolite SDG is readily detectable at a concentration that is within the range of plasma concentrations found in humans, although higher than the average level (Fig. 1). The concentration of the synthetic selenium derivative p-XSC required for Fas-L induction is higher than the concentration of SDG, but it is well established in animal models that p-XSC is less toxic and exerts its maximum cancer-protective effect at high dietary levels [~30 ppm compared with 2 ppm for selenite (83)].

Another recent finding is that induction of the so-called Jun NH$_2$-terminal kinase (JNK) and p38 stress kinases,

members of the mitogen-activated protein kinase family, is associated with induction of apoptosis by SDG and p-XSC (76). However, functional intervention experiments show that the JNK stress kinase pathway specifically is most important for induction of Fas-L by SDG (76) [but this is not the case for p-XSC, for reasons that remain unclear (76)]. There is other evidence in the literature also indicating that the JNK-c-Jun pathway is mechanistically upstream of Fas-L in other contexts, for example, after treatment with anticancer drugs or alkylating agents, but this is not necessarily the case (reviewed in Ref. 84). How selenium metabolites, such as SDG, might activate the JNK-Fas-L pathway is not clear, but one possibility, in view of our evidence that changes in redox balance may be important (Fig. 4), is that redox regulation of the upstream kinase (apoptosis signal-regulating kinase-1) that regulates JNK/p38 kinases (85) is involved, since apoptosis signal-regulating kinase-1 is known to be inhibited by reduced thioredoxin (86) and glutathione (87), and SDG has been shown to inhibit thioredoxin and thioredoxin reductase (58,88,89). We are currently attempting to test this hypothesis.

Acknowledgments and Notes

The authors thank Prof. J. Wyke for reading the manuscript, Dr. K. El-Bayoumy (Amercian Health Foundation, Valhalla, NY) for the gift of p-XSC, and Dr. S. Keyse (Biomedical Research Centre, Dundee, Scotland) for the gift of the heme oxygenase cDNA probe. This work was supported by the Cancer Research Campaign. Address correspondence to Paul R. Harrison, The Beatson Institute for Cancer Research, CRC Beatson Laboratories, Garscube Estate, Switchback Rd., Bearsden, Glasgow G61 1BD, Scotland, UK. E-mail: p.r.harrison@beatson.gla.ac.uk.

Submitted 28 March 2001; accepted in final form 25 May 2001.

References

1. Medina D: Mechanisms of selenium inhibition of tumorigenesis. *J Am Coll Toxicol* **5**, 21–27, 1986.

2. Poirier KA and Milner JA: Factors affecting the antitumorigenic properties of selenium in mice. *J Nutr* **113**, 2147–2154, 1983.

3. Thompson HJ, Meeker LD, and Kokosa AS: Effect of inorganic and organic forms of dietary selenium on the promotional phase of mammary carcinogenesis in the rat. *Cancer Res* **44**, 2803–2806, 1984.

4. Ip C: Lessons from basic research in selenium and cancer prevention. *J Nutr* **128**, 1845–1854, 1998.

5. Combs GF and Gray WP: Chemopreventive agents: selenium. *Pharmacol Ther* **79**, 179–192, 1998.

6. Ip C and Ganther HE: Activity of methylated forms of selenium in cancer prevention. *Cancer Res* **50**, 1206–1211, 1990.

7. Ip C, Hayes C, Budnick RM, and Ganther HE: Chemical form of selenium, critical metabolites, and cancer prevention. *Cancer Res* **51**, 595–600, 1991.

8. El-Bayoumy K: Effects of organoselenium compounds on induction of mouse forestomach tumors by benzo[a]pyrene. *Cancer Res* **45**, 3631–3635, 1985.

9. Reddy BS, Upadhyaya P, Simi B, and Rao CV: Evaluation of organoselenium compounds for potential chemopreventive properties in colon carcinogenesis. *Anticancer Res* **14**, 2509–2514, 1994.

10. Nayini J, El-Bayoumy K, Sugie S, Cohen LA, and Reddy BS: Chemoprevention of experimental mammary carcinogenesis by the synthetic organoselenium compound, benzylselenocyanate, in rats. *Carcinogenesis* **10**, 509–512, 1989.

11. Ip C, El-Bayoumy K, Upadhyaya P, Ganther HE, Vadhanavikit S, et al.: Comparative effect of inorganic and organic selenocyanate derivatives in mammary cancer chemoprevention. *Carcinogenesis* **15**, 187–192, 1994.

12. Ip C, Vadhanavikit S, and Ganther H: Cancer chemoprevention by aliphatic selenocyanates: effect of chain length on inhibition of mammary tumors and DMBA adducts. *Carcinogenesis* **16**, 35–38, 1995.

13. El-Bayoumy K, Upadhyaya P, Desai DH, Amin S, and Hecht SS: Inhibition of 4-(methylnitrosamino)-1-(3-pyridyl)-1-butanone tumorigenicity in mouse lung by the synthetic organoselenium compound, 1,4-phenylenebis(methylene)selenocyanate. *Carcinogenesis* **14**, 1111–1113, 1993.

14. Reddy BS, Sugie S, Maruyama H, Bayoumy K, and Marra P: Chemoprevention of colon carcinogenesis by the synthetic organoselenium compound, 1,4-phenylenebis(methylene)selenocyanate. *Cancer Res* **52**, 5635–5640, 1987.

15. El-Bayoumy K, Chae YH, Upadhyaya P, Meschter C, Cohen LA, et al.: Inhibition of 7,12-dimethylbenz[a]anthracene-induced tumors and DNA adduct formation in the mammary glands of female Spague-Dawley rats by the synthetic organoselenium compound, 1,4-phenylenebis(methylene)selenocyanate. *Cancer Res* **52**, 2402–2407, 1992.

16. Cai X-J, Block E, Uden PC, Zhang X, Quimby BD, et al.: *Allium* chemistry: identification of selenoaminoacids in ordinary and selenium-enriched garlic, onion and broccoli using gas chromatography with atomic emission detection. *J Agric Food Chem* **43**, 1754–1757, 1995.

17. Willett WC, Morris JS, Pressel S, Taylor JO, Polk BF, et al.: Pre-diagnostic serum selenium and risk of cancer. *Lancet* **2**, 130–134, 1983.

18. Salonen JT, Alfthan G, Huttunen JK, and Puska P: Association between serum selenium and the risk of cancer. *Am J Epidemiol* **120**, 342–349, 1984.

19. Peleg I, Morris S, and Hames CG: Is serum selenium a risk factor for cancer? *Med Oncol Tumor Pharmacother* **2**, 157–163, 1985.

20. Salonen JT, Salonen R, Lappettelainen R, Maenpaa PH, Alfthan G, et al.: Risk of cancer in relation to serum concentrations of selenium and vitamins A and E: matched case-control analysis of prospective data. *Br Med J* **290**, 417–420, 1985.

21. Fex G, Pettersson, B, and Akesson B: Low plasma selenium as a risk factor for cancer death in middle-aged men. *Nutr Cancer* **10**, 221–229, 1987.

22. Kok FJ, de Bruijn M, Hofman A, Vermeeren R, and Valkenburg HA: Is serum selenium a risk factor for cancer in men only? *Am J Epidemiol* **125**, 12–16, 1987.

23. Nomura A, Heilbrun LK, Morris JS, and Stemmermann GN: Serum selenium and risk of cancer, by specific sites: case-control analysis of prospective data. *JNCI* **79**, 103–108, 1987.

24. Virtamo J, Valkeila E, Alfthan U, Punsar S, Huttunen JK, et al.: Serum selenium and risk of cancer: a prospective follow-up of nine years. *Cancer* **60**, 145–148, 1987.

25. Coates RJ, Weiss NS, Daling JR, Morris JS, and Labbe RF: Serum levels of selenium and retinol and the subsequent risk of cancer. *Am J Epidemiol* **128**, 515–523, 1988.

26. Ringstad, J, Jacobsen BK, Tretli S, and Thommassen Y: Serum selenium concentration associated with risk of cancer. *J Clin Pathol* **41**, 454–457, 1988.

27. Knekt P, Aromaa A, Maatela J, Alfthan G, Aaran R-K, et al.: Serum selenium and subsequent risk of cancer among Finnish men and women. *JNCI* **82**, 864–868, 1990.

28. Garland M, Morris JS, Stampfer MJ, Colditz GA, Spate VL, et al.: A prospective study of toenail selenium levels and cancer among women. *JNCI* **87**, 497–505, 1995.

29. van den Brandt PA, Goldbohm RA, van't Veer P, Bode P, Dorant E, et al.: A prospective study on selenium status and the risk of lung cancer. *Cancer Res* **53**, 4860–4865, 1993.

30. Yoshikawa K, Willett WC, Morris SJ, Stampfer MJ, Spiegelman D, et al.: Study of prediagnostic selenium level in toenails and the risk of advanced prostate cancer. *JNCI* **90**, 1219–1224, 1998.

31. Menkes MS, Comstock GW, Vuilleumier JP, Helsing KJ, Rider AA, et al.: Serum β-carotene, vitamins A and E, selenium and the risk of lung cancer. *N Engl J Med* **315**, 1250–1254, 1986.

32. van den Brandt PA, Goldbohm RA, van't Veer P, Bode P, Dorant E, et al.: A prospective cohort study on toenail selenium levels and risk of gastrointestinal cancer. *JNCI* **85**, 224–229, 1993.

33. Mark SD, Qiao Y-L, Dawsey SM, Wu Y-P, Katki H, et al.: Prospective study of serum selenium levels and incident esophageal and gastric cancers. *JNCI* **92**, 1753–1763, 2000.

34. Hunter DJ, Morris JS, Stampfer MJ, Colditz GA, Speizer FE, et al.: A prospective study of selenium status and breast cancer risk. *JAMA* **264**, 1128–1131, 1990.

35. van den Brandt PA, Goldbohm RA, van't Veer P, Bode P, Dorant E, et al.: Toenail selenium levels and the risk of breast cancer. *Am J Epidemiol* **140**, 20–26, 1994.

36. Overvad K, Wang DY, Olsen J, Allen DS, Thorling EB, et al.: Selenium in human mammary carcinogenesis: a case-cohort study. *Eur J Cancer* **27**, 900–902, 1991.

37. Yu S-Y, Zhu Y-J, Li W-G, Huang Q-S, Zhi-Huang C, et al.: A preliminary report on the intervention trials of primary liver cancer in high-risk populations with nutritional supplementation of selenium in China. *Biol Trace Elem Res* **29**, 289–294, 1991.

38. Blot WJ, Li J-Y, Taylor PR, Guo W, Dawsey S, et al.: Nutrition intervention trials in Linxian, China: supplementation with specific vitamin/mineral combinations, cancer incidence and disease-specific mortality in the general population. *JNCI* **85**, 1483–1492, 1993.

39. Li J-Y, Taylor PR, Li B, Dawsey S, Wang G-Q, et al.: Nutrition intervention trials in Linxian, China: multiple vitamin/mineral supplemen-

tation, cancer incidence and disease-specific mortality among adults with esophageal dysplasia. *JNCI* **85**, 1492–1498, 1993.

40. Lee I-M, Cook NR, Manson JE, Buring JE, and Hennekens CH: β-Carotene supplementation and incidence of cancer and cardiovascular disease: The Women's Health Study. *JNCI* **91**, 2102–2106, 1999.

41. Greenberg ER, Baron JA, Tosteson TD, Freeman DH, Beck GJ, et al.: A clinical trial of antioxidant vitamins to prevent colorectal adenoma. *N Engl J Med* **331**, 141–147, 1994.

42. Hennekens CH, Buring JE, Manson JE, Stampfer M, Rosner B, et al.: Lack of effect of long-term supplementation with β-carotene on the incidence of malignant neoplasms and cardiovascular disease. *N Engl J Med* **334**, 1145–1149, 1996.

43. Lee I-M, Cook NR, Manson JE, Buring JE, and Hennekens H: β-Carotene supplementation and incidence of cancer and cardiovascular disease: The Women's Health Study. *JNCI* **91**, 2102–2106, 1999.

44. Hartmari TJ, Albanes D, Pietinen P, Hartman AM, Rautalahti M, et al.: The association between baseline vitamin E, selenium and prostate cancer in the α-Tocopherol, β-Carotene Cancer Prevention Study. *Cancer Epidemiol Biomarkers Prev* **7**, 335–340, 1998.

45. Heinonen OP, Albanes D, Virtamo J, Taylor PR, Huttunen JK, et al.: Prostate cancer and supplementation with α-tocopherol and β-carotene: incidence and mortality in a controlled trial. *JNCI* **18**, 440–446, 1998.

46. ATBC Cancer Prevention Study Group: The effect of vitamin E and β-carotene on the incidence of lung cancer and other cancers in male smokers. *N Engl J Med* **330**, 1029–1035, 1994.

47. Albanes D, Heinonen OP, Taylor PR, Virtamo J, Edwards BK, et al.: α-tocopherol and β-carotene supplements and lung cancer incidence in the α-Tocopherol, β-Carotene Cancer Prevention Study: effects of base-line characteristics and study compliance. *JNCI* **88**, 1560–1570, 1996.

48. Omenn GS, Goodman GE, Thornquist MD, Balmes J, Cullen MR, et al.: Effects of a combination of β-carotene and vitamin A on lung cancer and cardiovascular disease. *N Engl J Med* **334**, 1150–1155, 1996.

49. Clark LC, Combs GF, Turnbull BW, Slate EH, Chalker DK, et al.: Effects of selenium supplementation for cancer prevention in patients with carcinoma of the skin. *JAMA* **276**, 1957–1963, 1996.

50. Clark LC, Dalkin B, Krongrad A, Combs GF, Turnbull BW, et al.: Decreased incidence of prostate cancer with selenium supplementation: results of a double-blind cancer prevention trial. *Br J Urol* **81**, 730–734, 1998.

51. Burk RF, Hill KE, Read R, and Bellew T: Response of rat selenoprotein P to selenium administration and fate of its selenium. *Am J Physiol* **261**, E26–E30, 1991.

52. Weitzel F, Ursini F, and Wendel A: Phospholipid hydroperoxide glutathione peroxidase in various mouse organs during selnium deficiency and repletion. *Biochim Biophys Acta* **1036**, 88–94, 1990.

53. Bermano G, Nicol F, Dyer JA, Sunde RA, Beckett GJ, et al.: Tissue-specific regulation of selenoenzyme gene expression during selenium deficiency in rats. *Biochem J* **311**, 425–430, 1995.

54. Ho Y-S, Magnenat J-L, Bronson RT, Cao J, Gargano M, et al.: Mice deficient in cellular glutathione peroxidase develop normally and show no increased sensitivity to hyperoxia. *J Biol Chem* **272**, 16644–16651, 1997.

55. Cheng WH, Ho YS, Ross DA, Valentine BA, Combs GF, et al.: Cellular glutathione peroxidase knockout mice express normal levels of selenium-dependent plasma and phospholipid hydroperoxide glutathione peroxidases in various tissues. *J Nutr* **127**, 1445–1450, 1997.

56. de Haan JB, Bladier C, Griffiths P, Kelner M, O'Shea RD, et al.: Mice with a homozygous null mutation for the most abundant gluthione peroxidase, Gpx1, show increased susceptibility to the oxidative stress-inducing agents paraquat and hydrogen peroxide. *J Biol Chem* **273**, 22528–22536, 1998.

57. Lu Y-P, Lou Y-R, Yen P, Newmark HL, Mirochnitchenko OI, et al.: Enhanced skin carcinogenesis in transgenic mice with high expression of glutathione peroxidase or both glutathione peroxidase and superoxide dismutase. *Cancer Res* **57**, 1468–1474, 1997.

58. Ganther HE: Selenium metabolism, selenoproteins and mechansims of cancer prevention: complexities with thioredoxin reductase. *Carcinogenesis* **20**, 1657–1666, 1999.

59. Kryukov GV, Kryukov VM, and Glasyshev VM: New mammalian selenocysteine-containing proteins identified with an algorithm that searches for selenocysteine insertion sequences. *J Biol Chem* **274**, 33888–33897, 1999.

60. Kumaraswamy E, Malykh A, Korotkov KV, Kozyavkin S, Hu Y, et al.: Structure-expression relationships of the 15 kDa selenoprotein: possible role of the protein in cancer etiology. *J Biol Chem* **275**, 35540–35547, 2000.

61. Ejadi S, Bhattacharya I, Voss K, Singletary K, and Milner JA: In vitro and in vivo effects of sodium selenite on 7,12-dimethylbenz[a]anthracene-DNA adduct formation in isolated rat mammary epithelial cells. *Carcinogenesis* **10**, 823–826, 1989.

62. El-Bayoumy K, Chae Y-H, Upadhyaya P, Cohen LA, and Reddy BS: Inhibition of 7,12-dimethylbenz[a]anthracene-induced tumors and DNA adduct formation in the mammary glands of female Sprague-Dawley rats by the synthetic selenium compound, 1,4-phenylene-bis(methylene)selenocyanate. *Cancer Res* **52**, 2402–2407, 1992.

63. Fiala ES, Joseph C, Sohn O-S, El-Bayoumy K, and Reddy BS: Mechanism of benzylselenocyanate inhibition of azo-methane-induced colon carcinogenesis in F344 Rats. *Cancer Res* **51**, 735–741, 1991.

64. Wilson AC, Thompson HJ, Schedin PJ, Gibson NW, and Ganther HE: Effect of methylated forms of selenium on cell viability and the induction of DNA strand breakage. *Biochem Pharmacol* **43**, 1137–1141, 1992.

65. Lanfear J, Fleming J, Wu L, Webster G, and Harrison PR: The selenium metabolite selenodiglutathione induces p53 and apoptosis: relevance to the chemopreventive effects of selenium. *Carcinogenesis* **15**, 1387–1392, 1994.

66. Thompson HJ, Wolson A, Lu J, Singh M, Jiang C, et al.: Comparison of the effects of an organic and inorganic form of selenium on a mammary carcinoma cell line. *Carcinogenesis* **15**, 183–186, 1994.

67. Ronai Z, Tillotson JK, Traganos F, Darynkiewicz Z, Conaway CC, et al.: Effects of organic and inorganic selenium compounds on rat mammary tumor cells. *Int J Cancer* **63**, 428–434, 1995.

68. Lu J, Kaek M, Jiang C, Wilson A, and Thompson H: Selenite induction of DNA strand breaks and apoptosis in mouse leukemic L1210 cells. *Biochem Pharmacol* **47**, 1531–1535, 1994.

69. Kaeck M, Lu J, Strange R, Ip C, Ganther HE, et al.: Differential induction of growth arrest inducible genes by selenium compounds. *Biochem Pharmacol* **53**, 921–926, 1997.

70. Spallholz JE: On the nature of selenium toxicity and carcinostatic activity. *Free Radic Biol Med* **17**, 45–64, 1994.

71. Seko Y, Saito Y, Kitahara J, and Imura N: Active oxygen generation by the reaction of selenite with reduced glutathione in vitro. In *Selenium in Biology and Medicine*, Wendel A (ed). Berlin: Springer-Verlag, 1989 pp 70–73.

72. Yan L and Spallholz JE: Generation of reactive oxygen species from the reaction of selenium compounds with thiols and mammary tumor cells. *Biochem Pharmacol* **45**, 429–437, 1993.

73. Wu L, Lanfear J, and Harrison PR: The selenium metabolite selenodiglutathione induces cell death by a mechanism distinct from H_2O_2 toxicity. *Carcinogenesis* **16**, 1579–1584, 1995.

74. Keyse SM and Tyrrell RM: Heme oxygenase is the major 32-kDa stress protein induced in human skin fibroblasts by UVA radiation, hydrogen peroxide and sodium arsenite. *Proc Natl Acad Sci USA* **86**, 99–103, 1989.

75. Powis G, Mustacich D, and Coon A: The role of the redox protein thioredoxin in cell growth and cancer. *Free Radic Biol Med* **29**, 312–322, 2000.

76. Ghose A, Fleming J, El-Bayoumy K, and Harrison PR: Enhanced sensitivity of human oral carcinomas to induction of apoptosis by selenium compounds: involvement of MAP kinase and Fas pathways. *Cancer Res*. In press.

77. McGregor F, Wagner E, Felix D, Soutar D, Parkinson K, et al.: Inappropriate retinoic acid receptor-β expression in oral dysplasias: corre-

lation with acquisition with the immortal phenotype. *Cancer Res* **57**, 3886–3889, 1997.

78. Edington KG, Loughran OP, Berry IJ, and Parkinson EK: Cellular immortality: a late event in the progression of human squamous cell carcinoma of the head and neck associated with p53 alteration and a high frequency of allele loss. *Mol Carcinog* **13**, 254–265, 1995.

79. Menter DG, Sabichi AL, and Lippman SM: Selenium effects on prostate cell growth. *Cancer Epidemiol Biomarkers Prev* **9**, 1171–1182, 2000.

80. Gastman BR, Atarashi Y, Reichert TE, Saito T, Balkir L, et al.: Fas ligand is expressed on human squamous cell carcinomas of the head and neck, and it promotes apotosis of T lymphocytes. *Cancer Res* **59**, 5336–5364, 2000.

81. Krammer PH: CD95's deadly mission in the immune system. *Nature* **407**, 789–795, 2000.

82. Kiremidjian-Schumacher L, Roy M, Glickman R, Schneider K, Rothstein S, et al.: Selenium and immunocompetence in patients with head and neck cancer. *Biol Trace Elem Res* **73**, 97–111, 2000.

83. Tanaka T, Makita H, Kawabata K, Mori H, and El-Bayoumy K: 1,4-phenylenebis(methylene)selenocyanate exerts exceptional chemopreventive activity in rat tongue carcinogenesis. *Cancer Res* **57**, 3644–3648, 1997.

84. Davis RJ: Signal transduction by the JNK group of MAP kinases. *Cell* **103**, 239–252, 2000.

85. Ichijo H, Nishida E, Irie K, ten Dijke P, Saitoh M, et al.: Induction of apoptosis by ASK1, a mammalian MAPKKK that activates SAPK/JNK and p38 signaling pathways. *Science* **275**, 90–94, 1997.

86. Saitoh M, Nishitoh H, Fujii M, Takeda K, Tobiume K, et al.: Mammalian thioredoxin is a direct inhibitor of apoptosis signal-regulating kinase (ASK)1. *EMBO J* **17**, 2596–2606, 1998.

87. Wilhelm D, Bender K, Knebel A, and Angel P: The level of intracellular glutathione is a key regulator for the induction of stress-activated signal transduction pathways including Jun N-terminal protein kinases and p38 kinase by alkylating agents. *Mol Cell Biol* **17**, 4792–4800, 1997.

88. Bjornstedt M, Kumar S, and Holmgren A: Selenodiglutathione is a highly efficient oxidant of reduced thioredoxin and a substrate for mammalian thioredoxin reductase. *J Biol Chem* **267**, 8030–8034, 1992.

89. Kumar S, Bjornstedt M, and Holmgren A: Selenite is a substrate for calf thymus thioredoxin reductase and thioredoxin and elicits a large non-stoichiometric oxidation of NADPH in the presence of oxygen. *Eur J Biochem* **207**, 435–439, 1992.

90. Locknecker MP, Taylor PR, Levander OA, Howe SM, Veillon C, et al.: Selenium in diet, blood, and toenails in relation to human health in a seleniferous area. *Am J Clin Nutr* **53**, 1288–1294, 1991.

NUTRITION AND CANCER, *40*(1), 50–54

Molecular Targets for Selenium in Cancer Prevention

Y. S. Kim and J. Milner

Abstract: Mounting evidence reveals that selenium is a dietary constituent with anticarcinogenic and antitumorigenic properties. Various forms of selenium appear to be effective in bringing about these effects, although preclinical studies suggest that differences may arise as the quantity provided is reduced. The literature also documents the greater sensitivity of neoplastic cells to selenium than their nonneoplastic counterparts. Unfortunately, the minimal amount needed to bring about a positive effect in humans remains elusive. If there is a positive response to exaggerated intakes, it will likely be dependent on many factors, including the consumption of other dietary constituents, as well as variation in a host of genetic pathways involved with cancer. Although the biological basis of the reduction in cancer risk ascribed to selenium remains to be established, its consistency in retarding various experimentally induced tumors and suppressing the growth of various types of neoplasms in vitro and in vivo suggests that several mechanisms are involved. Depressed carcinogen bioactivation, reduced cell proliferation, and increased apoptosis raise the possibility that selenium works at a number of specific molecular targets involved with the cancer process. This review will focus on molecular targets involved with cell proliferation and apoptosis as possible mechanisms by which selenium might alter the cancer process.

Introduction

Selenium is an essential nutrient and, thus, has fundamental importance in maintaining health. For decades it was recognized for its ability to serve in conjunction with vitamin E to retard vascular and muscular dystrophy in animals (1). Early studies also identified its critical role in sperm formation and, thus, overall reproductive capacity (2). Its antioxidant properties stem from its key regulatory function within several selenoproteins (3). Today, ~20 eukaryotic and 15 prokaryotic selenoproteins containing selenocysteine have been identified, partially characterized, and/or cloned. Although considerable attention has been given to the potential health benefits of increasing selenium intakes, its importance as a modifier of cardiovascular disease risk remains

equivocal (4). Nevertheless, considerable evidence points to the importance of an adequate supply of selenium for maintaining immunocompetence (5). Likewise, it has been reported to have a fundamental role in determining the virulence of some viruses (6). As mentioned below, substantial evidence indicates that selenium may alter cancer at several sites and by multiple mechanisms.

Because carcinogen metabolism and immunocompetence are discussed elsewhere in this special issue, this review focuses on how selenium might influence molecular targets associated with cell proliferation and apoptosis. Some of the possible molecular targets that selenium may alter include nuclear factor-κB (NF-κB), activator protein-1 (AP-1), cdk2, cyclooxygenase, and/or lipoxygenase.

Anticancer Effects

Almost 30 years ago, evidence surfaced that selenium might be a physiologically important deterrent to cancer (7,8). Since that time, several epidemiological and preclinical studies have added to the belief that higher intakes of selenium might retard the incidence and biological behavior of a variety of tumors. The more recent study by Clark and associates (9) provides some of the most compelling data that selenium might truly be a deterrent to cancer. In their study, a 200-μg selenium supplement per day for a mean of 4.5 yr, as selenized yeast, depressed cancer-related mortality by ~40%. Additional clinical intervention studies are needed to confirm these observations.

The ability of selenium to retard chemically induced cancers, including those induced in mammary tissue, prostate, lung, colon, pancreas, and liver, suggests that a common metabolic change, rather than a tissue-specific reaction, may account for its physiological actions (10–12). However, changes in carcinogen metabolism cannot totally explain the anticancer effects of selenium, especially those associated with altered rates of neoplastic proliferation (13–18).

The impact of the form of selenium on the efficacy of cancer prevention has not been extensively examined. Nevertheless, several selenium-containing compounds with diverse chemical structures have been found to inhibit cell

The authors are affiliated with the Nutritional Science Research Group, Division of Cancer Prevention, National Cancer Institute, Rockville, MD 20892. J. Milner is also affiliated with the Nutrition Department, The Pennsylvania State University, University Park, PA 16803.

proliferation. Although limited evidence exists, organic selenium compounds such as selenomethionine and selenocysteine may be slightly less effective on a molar basis than selenite (13). The form of selenium was also important in a chemically induced lung tumor model when synthetic 1,4-phenylenebis(methylene)selenocyanate (p-XSC) was found to be superior to selenite (19). In a 7,12-dimethylbenz[a]anthracene (DMBA)-induced mammary model, p-XSC was again shown to be more effective than selenite in inhibiting tumors (20). In vitro studies by Thompson et al. (21) compared the effects of p-XSC and selenite on the induction of apoptosis in a mouse mammary carcinoma cell line. They found that p-XSC was far more effective than selenite. Exposure of mammary tumor cells to p-XSC has been found to suppress DNA, RNA, and protein synthesis, as well as inhibit mitochondrial transmembrane potential (22). Likewise, p-XSC has been shown to inhibit protein kinase C and protein kinase A activities in fibroblast cells (23). Although comparisons of selenite and p-XSC are interesting, variations in the biological response to other forms of selenium are less well documented. The use of exaggerated quantities of selenium in dietary and tissue culture studies makes it difficult to detect significant differences among forms.

Experimentally, the anticancer properties of selenium appear to arise at intakes that are substantially greater than those associated with maximal expression of most selenium-containing enzymes. Thus, although it is possible that one or more of the selenoenzymes might account for some of the anticancer protection, much of the evidence points to selenium metabolites as instrumental in the overall biological response (10,12).

Intracellularly Generated Selenocompounds

Historically, research has revealed the propensity of selenium to interact with thiols, especially protein thiols. Selenate is slowly reduced to selenite by glutathione and other sulfhydryl compounds. Selenite is further reduced to relatively stable compounds such as selenotrisulfide and selenopersulfide. Both may produce free radicals and ultimately lead to DNA damage or alter a host of thiols and, thus, alter several metabolic events. Kuchan and Milner (24) provided rather compelling evidence that intracellular concentrations of glutathione were instrumental in determining the ability of selenite to alter cellular proliferation. Thus it is conceivable that the various forms of selenium will not be equal in their efficacy but will be highly dependent on not only the quantity provided but how it is metabolized.

Selenite and selenodiglutathione (GSSeSG) are recognized as efficient oxidants of reduced thioredoxin and reduced thioredoxin reductase (TR) (25). Selenite and other redox-active selenocompounds are recognized to modify a host of cellular proteins, including the tumor promoter protein kinase C (26). By using phorbol ester-promoted JB6 epidermal cell transformation assay, Gopalakrishna et al. (26) found that selenite-, selenocystine-, and selenodigluta-thione-inactivated protein kinase C was reversed by treatment with thiol agents. Spallholz (27) indicates that several forms of selenium are capable of forming Se-S bonds with thiol-containing amino acids and/or proteins.

The formation of selenotrisulfide or selenenylsulfide bonds may be important in the anticancer properties associated with selenium. Several studies suggest that selenite may increase intracellular concentrations of the selenotrisulfide GSSeSG. Exposure of cells to GSSeSG has been found to markedly inhibit the growth of tumor cells, possibly by arresting protein biosynthesis (28–30). Although this exaggerated effect may relate to the delivery of selenium to a target site in the cell, it may also be one of the active forms of selenium that brings about a depression in tumor proliferation.

Monomethylated forms of selenium may also be involved in the chemopreventive effects of selenium (12,31). Selenite and selenomethionine exposures are known to increase the formation of methylated metabolites, including methylselenol, dimethylselenide, and trimethylselenonium. In vitro experiments reveal that methylseleninic acid is more potent than Se-methylselenocysteine in retarding cell proliferation and promoting apoptosis whether wild-type or nonfunctional p53 was present in hyperplastic mammary epithelial cells. Furthermore, the change in proliferation or apoptosis could not be attributed to DNA damage. In general, methylseleninic acid was ~10 times more effective than Se-methylselenocysteine in cells in culture (32). It is possible that cells vary in their ability to generate a monomethylated selenium species from Se-methylselenocysteine because of variation in the activity of β-lyase. The overall significance of this variation is unclear, since in vivo studies suggest that methylseleninic acid and Se-methylselenocysteine are comparable in retarding tumors resulting from methylnitrosourea or DMBA (32).

Selenium Inactivates NF-κB and AP-1

Although the mechanism by which selenium inhibited experimentally induced cancer remains to be determined, initial studies focused on selenoproteins such as cytoplasmic glutathione peroxidase (cGPx). In most preclinical studies, the protection provided by selenium occurred when it was provided at concentrations beyond that required to optimize the activity of cGPx and possibly other selenium-containing enzymes. Nevertheless, it is possible that a correction of cellular selenoenzyme deficiencies might account for some anticarcinogenic actions. Support for the involvement of several selenium-containing enzymes continues to emerge because of their involvement in determining the cell's redox status.

NF-κB is an inducible oncogenic nuclear transcription factor with a pivotal role in inducing genes involved in a number of physiological processes, including those associated with cytokines, growth factors, cell adhesion molecules, and immunoreceptors (33,34). The redox state is important in determining NF-κB activity. Support for this statement comes from the ability of chemically diverse antioxidants such as N-acetyl-L-cysteine, α-lipoic acid, butylated hydroxyanisole, pyrrolidine dithiocarbamate, and α-tocopherol to block its activation in vivo (35–37).

Several factors are known to influence the redox status of the cell, including the concentration of reactive oxygen species arising from lipid and H_2O_2. The modulation of NF-κB by specific reactive oxygen species appears to be cell specific, rather than a general phenomenon (38). Selenium and selenium-containing enzymes, including cGPx, membrane glutathione peroxidase, and TR, may also be involved via alterations in $H_2\dot{O}_2$ and lipid peroxide and/or thiol status (39–41). In human T47D cells, the overexpression of cGPx retards NF-κB activation, NF-κB nuclear translocation, and IκB degradation in response to tumor necrosis factor-α or H_2O_2 treatment. These phenomena were no longer observed when tetrameric cGPx activity was reduced by selenium depletion (40). Available evidence suggests that increased cGPx activity may interfere with the activation, but not the synthesis or stability, of NF-κB (40). Thus selenium intakes and variation in cGPx expression might be factors determining the rates of proliferation of some neoplasms.

Another glutathione peroxidase, phospholipid hydroperoxide glutathione peroxidase (PHGPx), is a monomeric, membrane-associated enzyme containing one atom of selenium per mole of protein. The involvement of this selenoenzyme in the interleukin-1 (IL-1)-induced NF-κB activation has been examined using the human umbilical endothelial cell line ECV 304 transfected with the PHGPx gene and the gene for selenophosphate synthase to foster selenoprotein biosynthesis (41). In these studies IL-1 induction of NF-κB was inhibited by supplementation with 50 nM selenite only in transfected cells. These results demonstrate that overexpression of PHGPx can inhibit NF-κB activation and suggests that NF-κB activation by IL-1 is mediated by a preferential substrate of PHGPx, such as a fatty acid hydroperoxide. Again, variation in selenium intake and PHGPx expression may account for some of the observed variation in rates of tumor proliferation.

The activity of TR, another selenoenzyme, has also been linked to NF-κB activation through its ability to regulate thioredoxin concentrations. Mammalian TR is homologous to glutathione reductase with a selenocysteine residue in the conserved COOH-terminal sequence (42). TR specifically reduces oxidized thioredoxin to its reduced form using NADPH (43). The reduced thioredoxin reduces disulfide bonds of several proteins, including NF-κB (44). Thioredoxin may associate with the NF-κB p50 subunit through its cysteine residues (45). In response to oxidative stress such as ultraviolet irradiation and tumor necrosis factor, human thioredoxin translocates from the cytoplasm into the nucleus, where it increases NF-κB transcriptional activity. The availability of selenium is a key factor that determines TR activity in cells in culture and in vivo (12,46). Providing supplemented selenium to HT-29 human colon cancer cells grown in serum-free medium markedly increased TR activity (47). Because feeding rats a high-selenium diet (1.0 ppm) has been reported to cause a transient increase in liver, kidney, and lung TR activity (46), the importance of this enzyme in explaining the antitumorigenic effects of selenium remains unclear.

AP-1 is another transcription factor involved in cell proliferation. Several forms of selenium are known to retard the binding of the AP-1 to DNA, presumably by altering redox control mechanisms. GSSeSG has been reported to be ~10 times more effective in inhibiting AP-1 DNA binding in nuclear extracts from 3B6 lymphocytes than is selenite (25). This nuclear transcription factor, as well as NF-κB, may be involved in the antitumorigenic effects of selenium.

Selenium Inhibits cdk2 and gadd45

Various forms of selenium markedly retard the growth of neoplasms. Part of this effect may relate to the recognized ability of selenite to produce DNA breakage and cell death (17,29,30). The induction of apoptosis has been attributed to changes in genes such as cyclin-dependent kinase 2 (cdk2) and gadd45 (48,49). The cdk2 and DNA damage-inducible (gadd) genes are related to cell cycle arrest at G_1/S and G_2/M, respectively. Genomic stability in eukaryotes can be maintained by the checkpoints at G_1/S and G_2/M in response to DNA damage. In vitro, methylselenocysteine has been reported to arrest mouse mammary tumor epithelial cells (TM6) in the S phase, which coincided with a specific block of cdk2 kinase activity and an elevated expression of gadd34, gadd45, and gadd153 (48,50). Although the underlying mechanism accounting for these observations is not clear, the alterations in cdk2 and gadd45 suggest that the effect of selenium in these cells may be related to the p53-mediated apoptosis. The p53 protein is a factor that enhances transcription of several genes, including gadd45, p21[WAFl, Cip-l], mdm2, cyclin G, bax, and insulin-like growth factor binding protein-3. Generally, the p53 protein is maintained at a low concentration, although it can be induced by physical or chemical DNA damage (51). p53 has been implicated in a G_2/M phase checkpoint, preventing premature entry into another S phase, possibly by altering gadd45 (52,53) and/or regulating the number of centrosomes in a cell (54).

Selenium and Apoptosis

Virtually all cells are endowed with the capacity for programmed cell death, i.e., apoptosis (55). This process typically involves activation of caspase-family cell death proteases. Apoptosis enhances the elimination of damaged and dysfunctional cells that may arise from oxidative stress, glycation, and DNA damage. Interestingly, apoptosis also can be triggered by selenium independent of DNA damage and in cells with a null p53 phenotype. p73 and p63 have homology to p53 in their respective transactivation, DNA-binding, and oligomerization domains. Both p73 and p63 transactivate p53-regulated promoters and induce apoptosis. Evidence suggests that p73 and p63 mediate apoptosis by mechanisms different from p53 (56). Although different forms of selenium have different effects on apoptosis (57–59), it remains to be determined whether selenium alters these or other factors associated with non-p53-mediated apoptosis.

Selenium Silences Lipoxygenases

PHGPx is an enzyme recognized for its involvement in the removal of esterified lipid hydroperoxides. Additionally, PHGPx may be involved with the silencing of several lipoxygenases, including 5-, 12-, and 15-lipoxygenase (60–62). The importance of this regulation stems from the recognition that lipoxygenases generate metabolites that mediate signals for increasing cell growth and proliferation (63) and inhibiting apoptosis (64). Creation of selenium-deficient rat basophilic leukemia cells with <1% of normal cGPx activity and ~35% of normal PHGPx activity caused an approximately eightfold increase in release of lipoxygenase metabolites compared with controls. Addition of 0.25 µg of selenium per milliliter of medium to these cells reduced the amount of 5-lipoxygenase metabolites released to control values after 12 h and restored PHGPx. Injection of 500 µg of selenium as Na_2SeO_3 per kilogram to rats raised leukocyte PHGPx activity eightfold and significantly decreased lipoxygenase-generated metabolites within 114 h compared with controls (60). These results indicate that PHGPx, but not cGPX, is likely responsible for silencing 5-lipoxygenase activity by regulating the tone of membrane hydroperoxides (65). The significance of the interactions between dietary fats and selenium is exemplified by the general association of a high dietary fat intake with increased risk of some cancers, such as prostate, and the protection that may occur with dietary selenium supplementation (66). Because the selenoprotein PHGPx can silence lipoxygenases, this may partially explain the observed anticancerous effects of this trace element.

Summary

Overall compelling evidence exists that selenium is a deterrent to cancer cell proliferation in model systems. Nevertheless, the best dietary source, quantity, and biologically active form of selenium remain to be determined. Likewise, there remains a dearth of information about the impact of dietary selenium on the biological behavior of tumors occurring in humans. It is likely that not all individuals will respond identically because of differences in their absorption and metabolism of selenocompounds. Genetic differences may also contribute to variation in response to selenium. Only by having knowledge of molecular targets for selenium can individuals be adequately identified who might benefit most or be placed at risk by a dietary strategy to enhance selenium intakes for cancer prevention.

Acknowledgments and Notes

Address correspondence to John Milner, Div. of Cancer Prevention, National Cancer Institute, 6130 Executive Bl., Rm. 212, Rockville, MD 20892. E-mail: milnerj@mail.nih.gov.

Submitted 7 December 2000; accepted in final form 31 January 2001.

References

1. Schwarz K, Bieri JG, Briggs GM, and Scott ML: Prevention of exudative diathesis in chicks by factor 3 and selenium. *Proc Soc Exp Biol Med* **95**, 621–625, 1957.
2. Hansen JC and Deguchi Y: Selenium and fertility in animals and man—a review. *Acta Vet Scand* **37**, 19–30, 1996.
3. Holben DH and Smith AM: The diverse role of selenium within selenoproteins: a review. *J Am Diet Assoc* **99**, 836–843, 1999.
4. Huttunen JK: Selenium and cardiovascular diseases—an update. *Biomed Environ Sci* **10**, 220–226, 1997.
5. Hughes DA: Effects of dietary antioxidants on the immune function of middle-aged adults. *Proc Nutr Soc* **58**, 79–84, 1999.
6. Beck MA and Levander OA: Dietary oxidative stress and the potentiation of viral infection. *Annu Rev Nutr* **18**, 93–116, 1998.
7. Shamberger RJ and Frost DV: Possible protective effect of selenium against human cancer. *Can Med Assoc J* **100**, 682, 1969.
8. Schrauzer GN and Rhead WJ: Interpretation of the methylene blue reduction test of human plasma and the possible cancer-protecting effect of selenium. *Experientia* **27**, 1069–1071, 1971.
9. Clark LC, Combs GF Jr, Turnbull BW, Slate EH, Chalker DK, et al.: Effects of selenium supplementation for cancer prevention in patients with carcinoma of the skin: a randomized controlled trial. Nutritional Prevention of Cancer Study Group. *JAMA* **276**, 1957–1963, 1996.
10. Milner JA: Inhibition of chemical carcinogenesis and tumorigenesis by selenium. *Adv Exp Med Biol* **206**, 449–463, 1986.
11. Schrauzer GN: Selenium: mechanistic aspects of anticarcinogenic action. *Biol Trace Elem Res* **33**, 51–62, 1992.
12. Ganther HE: Selenium metabolism, selenoproteins and mechanisms of cancer prevention: complexities with thioredoxin reductase. *Carcinogenesis* **20**, 1657–1666, 1999.
13. Greeder GA and Milner JA: Factors influencing the inhibitory effect of selenium on mice inoculated with Ehrlich ascites tumor cells. *Science* **209**, 825–827, 1980.
14. Milner JA and Hsu CY: Inhibitory effects of selenium on the growth of L1210 leukemic cells. *Cancer Res* **41**, 1652–1656, 1981.
15. Watrach AM, Milner JA, Watrach MA, and Poirier KA: Inhibition of human breast cancer cells by selenium. *Cancer Lett* **25**, 41–47, 1984.
16. Nano JL, Czerucka D, Menguy F, and Rampal P: Effect of selenium on the growth of three human colon cancer cell lines. *Biol Trace Elem Res* **20**, 31–43, 1989.
17. Lu J, Jiang C, Kaeck M, Ganther H, Vadhanavikit S, et al.: Dissociation of the genotoxic and growth inhibitory effects of selenium. *Biochem Pharmacol* **50**, 213–219, 1995.
18. Redman C, Scott JA, Baines AT, Basye JL, Clark LC, et al.: Inhibitory effect of selenomethionine on the growth of three selected human tumor cell lines. *Cancer Lett* **125**, 103–110, 1998.
19. el-Bayoumy K, Upadhyaya P, Desai DH, Amin S, and Hecht SS: Inhibition of 4-(methylnitrosamino)-1-(3-pyridyl)-1-butanone tumorigenicity in mouse lung by the synthetic organoselenium compound, 1,4-phenylenebis(methylene)selenocyanate. *Carcinogenesis* **14**, 1111–1113, 1993.
20. Ip C, el-Bayoumy K, Upadhyaya P, Ganther H, Vadhanavikit S, et al.: Comparative effect of inorganic and organic selenocyanate derivatives in mammary cancer chemoprevention. *Carcinogenesis* **15**, 187–192, 1994.
21. Thompson HJ, Wilson A, Lu J, Singh M, Jiang C, et al.: Comparison of the effects of an organic and an inorganic form of selenium on a mammary carcinoma cell line. *Carcnogenesis* **15**, 183–186, 1994.
22. Ronai Z, Tillotson JK, Traganos F, Darzynkiewicz Z, Conaway CC, et al.: Effects of organic and inorganic selenium compounds on rat mammary tumor cells. *Int J Cancer* **63**, 428–434, 1995.
23. Foiles PG, Fujiki H, Suganuma M, Okabe S, Yatsunami J, et al.: Inhibition of PKC and PKA by chemopreventive organoselenium compounds. *Int J Oncol* **7**, 685–690, 1995.

24. Kuchan MJ and Milner JA: Influence of intracellular glutathione on selenite-mediated growth inhibition of canine mammary tumor cells. *Cancer Res* **52**, 1091–1095, 1992.

25. Spyrou G, Bjornstedt M, Kumar S, and Holmgren A: AP-1 DNA-binding activity is inhibited by selenite and selenodiglutathione. *FEBS Lett* **368**, 59–63, 1995.

26. Gopalakrishna R, Chen ZH, and Gundimeda U: Selenocompounds induce a redox modulation of protein kinase C in the cell, compartmentally independent from cytosolic glutathione: its role in inhibition of tumor promotion. *Arch Biochem Biophys* **348**, 37–48, 1997.

27. Spallholz JE: Free radical generation by selenium compounds and their prooxidant toxicity. *Biomed Environ Sci* **10**, 260–270, 1997.

28. Vernie LN, Collard JG, Eker AP, de Wildt A, and Wilders IT: Studies on the inhibition of protein synthesis by selenodiglutathione. *Biochem J* **180**, 213–218, 1979.

29. Poirier KA and Milner JA: Factors influencing the antitumorigenic properties of selenium in mice. *J Nutr* **113**, 2147–2154, 1983.

30. Cho DY, Jung U, and Chung AS: Induction of apoptosis by selenite and selenodiglutathione in HL-60 cells: correlation with cytotoxicity. *Biochem Mol Biol Int* **47**, 781–793, 1999.

31. Ip C, Hayes C, Budnick RM, and Ganther HE: Chemical form of selenium, critical metabolites, and cancer prevention. *Cancer Res* **51**, 595–600, 1991.

32. Ip C, Thompson HJ, Zhu Z, and Ganther HE: In vitro and in vivo studies of methylseleninic acid: evidence that a monomethylated selenium metabolite is critical for cancer chemoprevention. *Cancer Res* **60**, 2882–2886, 2000.

33. Brown K, Gerstberger S, Carlson L, Franzoso G, and Siebenlist U: Control of Iκ B-α proteolysis by site-specific, signal-induced phosphorylation. *Science* **267**, 1485–1488, 1995.

34. Baldwin AS Jr: The NF-κ B and IκB proteins: new discoveries and insights. *Annu Rev Immunol* **14**, 649–683, 1996.

35. Baeuerle PA and Henkel T: Function and activation of NF-κB in the immune system. *Annu Rev Immunol* **12**, 141–179, 1994.

36. Flohe L, Brigelius-Flohe R, Saliou C, Traber MG, and Packer L: Redox regulation of NF-κB activation. *Free Radic Biol Med* **22**, 1115–1126, 1997.

37. Hsu TC, Young MR, Cmarik J, and Colburn NH: Activator protein 1 (AP-1)- and nuclear factor-κB (NF-κB)-dependent transcriptional events in carcinogenesis. *Free Radic Biol Med* **28**, 1338–1348, 2000.

38. Bowie A and O'Neill LA: Oxidative stress and nuclear factor-κB activation: a reassessment of the evidence in the light of recent discoveries. *Biochem Pharmacol* **59**, 13–23, 2000.

39. Tolando R, Jovanovic A, Brigelius-Flohe R, Ursini F, and Maiorino M: Reactive oxygen species and proinflammatory cytokine signaling in endothelial cells: effect of selenium supplementation. *Free Radic Biol Med* **28**, 979–986, 2000.

40. Kretz-Remy C, Mehlen P, Mirault ME, and Arrigo AP: Inhibition of IκB-α phosphorylation and degradation and subsequent NF-κB activation by glutathione peroxidase overexpression. *J Cell Biol* **133**, 1083–1093, 1996.

41. Brigelius-Flohe R, Friedrichs B, Maurer S, Schultz M, and Streicher R: Interleukin-1-induced nuclear factor-κB activation is inhibited by overexpression of phospholipid hydroperoxide glutathione peroxidase in a human endothelial cell line. *Biochem J* **328**, 199–203, 1997.

42. Zhong L and Holmgren A: Essential role of selenium in the catalytic activities of mammalian thioredoxin reductase revealed by characterization of recombinant enzymes with selenocysteine mutations. *J Biol Chem* **275**, 18121–18128, 2000.

43. Holmgren A and Bjornstedt M: Thioredoxin and thioredoxin reductase. *Methods Enzymol* **252**, 199–208, 1995.

44. Sun Q, Wu Y, Zappacosta F, Jeang K, Lee BJ, et al.: Redox regulation of cell signaling by selenocysteine in mammalian thioredoxin reductases. *J Biol Chem* **274**, 24522–24530, 1999.

45. Matthews JR, Wakasugi N, Virelizier JL, Yodoi J, and Hay RT: Thioredoxin regulates the DNA binding activity of NF-κB by reduction of a disulphide bond involving cysteine 62. *Nucleic Acids Res* **20**, 3821–3830, 1992.

46. Berggren MM, Mangin JF, Gasdaka JR, and Powis G: Effect of selenium on rat thioredoxin reductase activity: increase by supranutritional selenium and decrease by selenium deficiency. *Biochem Pharmacol* **57**, 187–193, 1999.

47. Berggren M, Gallegos A, Gasdaska J, and Powis G: Cellular thioredoxin reductase activity is regulated by selenium. *Anticancer Res* **17**, 3377–3380, 1997.

48. Sinha R, Kiley SC, Lu JX, Thompson HJ, Moraes R, et al.: Effects of methylselenocysteine on PKC activity, cdk2 phosphorylation and gadd gene expression in synchronized mouse mammary epithelial tumor cells. *Cancer Lett* **146**, 135–145, 1999.

49. Kaeck M, Lu J, Strange R, Ip C, Ganther HE, et al.: Differential induction of growth arrest inducible genes by selenium compounds. *Biochem Pharmacol* **53**, 921–926, 1997.

50. Sinha R and Medina D: Inhibition of cdk2 kinase activity by methylselenocysteine in synchronized mouse mammary epithelial tumor cells. *Carcinogenesis* **18**, 1541–1547, 1997.

51. Levine AJ: p53, the cellular gatekeeper for growth and division. *Cell* **88**, 323–331, 1997.

52. Cross SM, Sanchez CA, Morgan CA, Schimke MK, Ramel S, et al.: A p53-dependent mouse spindle checkpoint. *Science* **267**, 1353–1356, 1995.

53. Fornace AJ Jr and Harris CC: GADD45 induction of a G_2/M cell cycle checkpoint. *Proc Natl Acad Sci USA* **96**, 3706–3711, 1999.

54. Fukasawa K, Choi T, Kuriyama R, Rulong S, and Vande Woude GF: Abnormal centrosome amplification in the absence of p53. *Science* **271**, 1744–1747, 1996.

55. Kaufmann SH and Gores GJ: Apoptosis in cancer: cause and cure. *Bioessays* **22**, 1007–1017, 2000.

56. Sheikh MS and Fornace AJ Jr: Role of p53 family members in apoptosis. *J Cell Physiol* **182**, 171–181, 2000.

57. Sundaram N, Pahwa AK, Ard MD, Lin N, Perkins E, et al.: Selenium causes growth inhibition and apoptosis in human brain tumor cell lines. *J Neurooncol* **46**, 125–133, 2000.

58. Zhu Z, Jiang W, Ganther HE, Ip C, and Thompson HJ: In vitro effects of Se-allylselenocysteine and Se-propylselenocysteine on cell growth, DNA integrity, and apoptosis. *Biochem Pharmacol* **60**, 1467–1473, 2000.

59. Shen H, Yang C, Liu J, and Ong C: Dual role of glutathione in selenite-induced oxidative stress and apoptosis in human hepatoma cells. *Free Radic Biol Med* **28**, 1115–1124, 2000.

60. Brigelius-Flohe R: Tissue-specific functions of individual glutathione peroxidases. *Free Radic Biol Med* **27**, 951–965, 1999.

61. Chen CJ, Huang HS, Lin SB, and Chang WC: Regulation of cyclooxygenase and 12-lipoxygenase catalysis by phospholipid hydroperoxide glutathione peroxidase in A431 cells. *Prostaglandins Leukot Essent Fatty Acids* **62**, 261–268, 2000.

62. Schnurr K, Brinckmann R, and Kuhn H: Cytokine-induced regulation of 15-lipoxygenase and phospholipid hydroperoxide glutathione peroxidase in mammalian cells. *Adv Exp Med Biol* **469**, 75–81, 1999.

63. Rao GN, Baas AS, Glasgow WC, Eling TE, Runge MS, et al.: Activation of mitogen-activated protein kinases by arachidonic acid and its metabolites in vascular smooth muscle cells. *J Biol Chem* **269**, 32586–32591, 1994.

64. Ghosh J and Myers CE: Inhibition of arachidonate 5-lipoxygenase triggers massive apoptosis in human prostate cancer cells. *Proc Natl Acad Sci USA* **95**, 13182–13187, 1998.

65. Weitzel F and Wendel A: Selenoenzymes regulate the activity of leukocyte 5-lipoxygenase via the peroxide tone. *J Biol Chem* **268**, 6288–6292, 1993.

66. Whittemore AS, Kolonel LN, Wu AH, John EM, Gallagher RP, et al.: Prostate cancer in relation to diet, physical activity, and body size in blacks, whites, and Asians in the United States and Canada. *JNCI* **87**, 652–661, 1995.

NUTRITION AND CANCER, *40*(1), 55–63

Protein Kinase C as a Molecular Target for Cancer Prevention by Selenocompounds

Rayudu Gopalakrishna and Usha Gundimeda

Abstract: Selenium is a very effective cancer-preventive agent, suppressing tumor promotion and early stages of tumor progression. However, the mechanisms by which selenium exerts these cancer-preventive actions are not known. Protein kinase C (PKC) is a receptor for certain tumor promoters and also plays a crucial role in events related to tumor progression. Therefore, it is not only a potential target for the cancer-preventive activity of selenium, but also it has the structural basis for interaction with selenium. Redox-active selenocompounds can inactivate PKC, particularly the Ca^{2+}-dependent isozymes, by reacting with the critical cysteine-rich regions present within the catalytic domain while, in some cases, also reacting with the cysteine residues present within the zinc-fingers of the regulatory domain. The selenoprotein thioredoxin reductase (TR), acting through thioredoxin, reverses the inactivation of PKC induced by selenometabolites. Furthermore, TR, through a direct interaction involving its selenosulfur center with the zinc-thiolates of PKC, can reverse the redox modification of this kinase induced by selenometabolites. Thus the selenometabolite-induced toxicity is reversed by a selenoprotein, and therefore an interrelationship exists between these two mechanisms of selenium actions. Moreover, this also explains how a resistance to selenium develops in advanced tumor cells probably due to an overexpression of functional TR. Selenium-induced inactivation of PKC may, at least in part, be responsible for the selenium-induced inhibition of tumor promotion, cell growth, invasion, and metastasis, as well as for the induction of apoptosis.

Introduction

Epidemiological data suggest that cancer mortality is inversely correlated with selenium consumption (1,2). Experimental studies in animals indicate that selenium supplementation at levels (1–3 ppm) above the dietary requirement (0.1 ppm) can prevent tumorigenesis at various sites, including breast and skin (3–12). Cancer prevention clinical trials carried out by Dr. Larry Clark and his associates suggested

that supplemental selenium may reduce the incidence and mortality of prostate, lung, and colorectal cancers, but not skin or breast cancer (13–15). This landmark achievement is not only a tribute to Dr. Larry Clark, but it also gives hope to cancer prevention researchers that prevention or delay of human cancer will eventually be possible through chemoprevention.

Many important issues remained to be answered to achieve cancer prevention in humans by selenium supplementation. Despite the well-demonstrated ability of selenium to inhibit breast cancer in animal studies (7–10), it is not clear whether selenium can decrease breast cancer in humans (13). Therefore, it is imperative to know why selenium prevents cancers in some cases while it fails to do so in other cases. Furthermore, it is important to understand whether a particular stage(s) in carcinogenesis is well suited for the cancer-preventive activity of selenium. Thus it is important to identify the molecular targets and mechanisms by which selenium prevents cancer. This will help us understand which types of cancers are best prevented by selenium and identify those individuals who would most benefit from selenium supplementation.

Selenium Inhibits Initiation and Postinitiation Stages in Carcinogenesis

To be most effective, chemopreventive agents should inhibit several stages in multistage carcinogenesis (16–18). Selenium, which can inhibit various stages in carcinogenesis, is thus well suited for prevention of carcinogenesis in humans. Selenium exerts its chemopreventive effects at the initiation phase of tumorigenesis by inhibiting carcinogen binding to DNA (19). In addition, selenium inhibits tumor promotion, progression, and angiogenesis (20–26). Selenium also stimulates the immune system (27). Experimental studies reveal that selenium is very effective at the postinitiation stages in carcinogenesis (10,28). The decades of tumor preneoplasia and/or early neoplasia, comprising tumor promotion or the early stages of tumor progression in

---

The authors are affiliated with the Department of Cell and Neurobiology, Keck School of Medicine, University of Southern California, Los Angeles, CA 90033.

the development of cancer in organs such as the prostate in humans, make these stages ideal for chemoprevention by selenium (29–31). However, the mechanisms by which selenium inhibits tumor promotion or progression are not known.

Selenium was shown to inhibit cell transformation in vitro at concentrations lower than that needed for the inhibition of growth of malignant cells (20–23). Furthermore, there is limited evidence to support the theory that dietary selenium can prevent the growth of transplanted tumors in animals (32–34). Recent in vitro studies have shown that various selenocompounds, especially at higher doses, induce apoptosis of tumor cells in culture (35–39). These growth-inhibiting and apoptosis-inducing mechanisms may be important in the promotion phase of carcinogenesis, where there is a clonal expansion of preneoplastic cells that escape death.

Selenium Actions as Selenoproteins and Selenometabolites

The original suggestion that selenium acts as a redox catalyst is an important clue to understand the actions of selenium present in trace amounts in the biological systems (40). It is believed that selenium can exert its cancer-preventive actions by two different mechanisms: one involves the actions of selenoproteins, and the other involves direct actions of selenometabolites. Both mechanisms have been found to have limitations; nevertheless, it is important to understand whether these two distinct mechanisms are interrelated and complementary.

Selenoproteins

At a nutritional level (0.1 ppm), selenium is essential to glutathione peroxidase (GPx), which protects cells from H_2O_2 and organic peroxides (41,42). Because peroxides can induce tumor promotion, there is a possibility that GPx may have a role in inhibiting tumor promotion. However, the requirement of selenium for cancer prevention (1–3 ppm), well above levels required for optimal expression of GPx activity, suggests involvement of complementary mechanisms for chemoprevention (8). Recently, selenocysteine was identified as the penultimate COOH-terminal residue in thioredoxin reductase (TR) (43–45). Unlike other selenoproteins previously characterized, TR activity is not saturated with nutritionally adequate dietary selenium, but a twofold increase in TR activity by increasing dietary selenium to cancer-preventive doses has been reported (46–49). Compared with normal tissues, tumors have increased expression of TR and thioredoxin, which are believed to protect cells from cell death (50). However, there are also suggestions that, under certain conditions, TR and thioredoxin can induce cell death, but the mechanism of this opposite action is unknown (51,52). Given the importance of oxidative stress in tumor promotion, the antioxidant functions of GPx and TR are very

relevant to the cancer-preventive actions of selenium. Nevertheless, their precise role is not clear.

Selenometabolites

Selenocompounds such as selenite, selenodiglutathione, selenomethionine, and Se-methylselenocysteine ultimately generate selenide (8), which is incorporated into the selenoproteins by a specialized mechanism (41,42). When selenide is generated in greater amounts, it also reacts with oxygen to produce superoxide and, ultimately, H_2O_2, which results in toxicity (53–55). This oxidative injury caused by selenide redox cycling is believed to be responsible for the toxic or cancer-preventive actions of selenium (53). Furthermore, selenide is methylated to methylselenol, dimethylselenide, and trimethylselenonium (4). Trimethylselenonium is excreted through urine, whereas dimethylselenide is exhaled (8). Although this methylation pathway was originally thought to be a detoxification pathway, previous elegant studies suggested a role for methylselenol and dimethylselenide in mediating the anticarcinogenic actions of selenium (8).

Although selenite, methylselenol, or other redox-active selenometabolites can induce certain cancer-preventive actions, they are present in very low concentrations in plasma and tissues. Several in vitro studies have shown growth inhibition of established malignant cells by selenite only at high (5–100 µM) concentrations (56–59). However, in vivo studies reveal that administered selenite is metabolized and incorporated into selenoproteins (60–62). Similarly, although selenomethionine inhibits tumor cell growth or induces apoptosis in vitro at high (40–130 µM) concentrations (63,64), it is present in plasma and tissues primarily in proteins, incorporated in place of methionine (65,66). Thus, at cancer-preventive doses, while excess selenium is excreted, most (>90%) selenium in the circulation is present as selenoproteins, and only a limited amount (<5%) is present as selenometabolites (67). Therefore, although these in vitro cellular actions of selenometabolites may be relevant to cancer prevention, it is important to determine whether these actions occur in vivo at bioavailable concentrations of selenium, as well as how they are related to the actions of selenoproteins. Therefore, a true molecular target for cancer-preventive actions of selenium may clarify the interrelationship between the actions of selenoproteins and selenometabolites.

Protein Kinase C as a Target for Tumor Promoters and Antitumor Promoters

Protein kinase C (PKC), a family of isozymes, is activated not only by lipid second messengers (68–70), but also by tumor promoters such as phorbol esters and oxidants (71–75). PKC regulates tumor promotion and cell growth by inducing activation of transcriptional factors, such as activator protein-1 (AP-1) and nuclear factor-κB (NF-κB), and by increasing the expression of key enzymes, such as ornithine decarboxylase, inducible nitric oxide synthase, and cyclooxygenase-2 (76–80).

PKC has unique structural aspects that render it susceptible to activation by oxidant tumor promoters, such as H_2O_2, periodate, and tobacco-related tumor promoters (71–73). Selective oxidative modification of the regulatory domain results in Ca^{2+}/lipid-independent activation, while selective oxidative modification of the kinase domain results in inactivation (81,82). The regulatory domain contains 12 cysteine residues that coordinate the binding of 4 zinc atoms; the zinc-thiolate structure is required for binding of phorbol ester and diacylglycerol (83,84). This positively charged zinc-thiolate is more susceptible to tumor-promoting oxidants (81,82). Thus phorbol esters activate the enzyme by binding to the structure supported by zinc-fingers, while oxidants directly induce a similar effect by reacting with zinc-thiolates and induce a collapse of the zinc-fingers (Fig. 1). In both cases, changes occurring in the regulatory domain relieve its autoinhibitory effect, caused by the interaction of its pseudosubstrate region with the protein substrate-binding region of the catalytic domain (C4). Furthermore, PKC is also activated by fatty acid hydroperoxides and other oxidation products (85–87), which is due to direct binding, similar to that of fatty acids.

Some cancer-preventive agents, such as polyphenolics (curcumin, 4-hydroxytamoxifen, and ellagic acid) in their oxidized state, can inactivate PKC by oxidizing the vicinal thiols present within the catalytic domain (88–90). Therefore, PKC, by having different oxidation-susceptible regions in the regulatory and catalytic domains, can respond to oxidant tumor promoters and chemopreventive agents to elicit opposite cellular responses. In addition to scavenging free radicals, antioxidants also interrupt signaling mechanisms triggered by oxidants (91,92). Although tumor promoters have their own receptors, it is not known whether chemopreventive agents also have their own receptors to induce their cellular actions. However, if they act on the same cellular target(s) as the tumor promoters, they can efficiently block cell signaling induced by tumor promoters.

Inactivation of PKC by Redox-Active Selenometabolites

Selenite

Several studies have explored the effects of selenium on PKC (93,94). At lower concentrations, selenite decreased PKC activity [concentration resulting in half-maximal inhibition (IC_{50}) = 0.5 μM] and induced a modification of four cysteine residues, resulting in the formation of two disulfides (93). However, at higher concentrations, it decreased phorbol ester binding and induced a modification of seven to eight cysteine residues, resulting in the formation of three to four disulfides. The isozymes α, β, and γ exhibited higher sensitivity to selenite than the ζ and δ isozymes. A cluster of at least four cysteine residues is needed for the rapid reaction of selenite with protein sulfhydryls (95). The four conserved cysteine residues present within the catalytic domain of α, β, and γ isozymes, although separated in the sequence, may be clustered in the tertiary structure, providing high specificity for selenite reaction with these PKC isozymes (93). However, PKC-ζ, which has only two of four of these conserved cysteines, reacted weakly with selenite; PKC-ε, which has only three conserved cysteine residues, was intermediate in exhibiting the sensitivity to selenite-induced inhibition (Fig. 2).

In intact cells, PKC modification by selenite was not interfered by glutathione, probably because of the shielding of the cysteine-rich region of the enzyme by a weak hydropho-

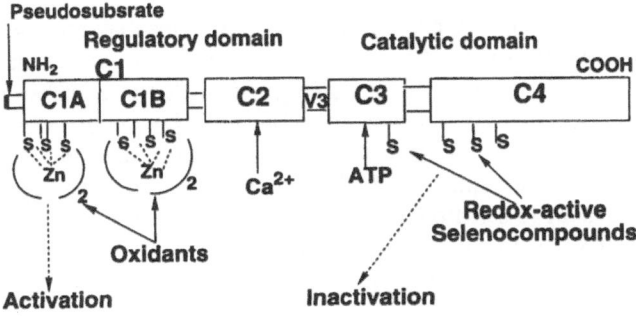

Figure 1. Multidomain structure of protein kinase C (PKC). Oxidation-susceptible sites within regulatory and catalytic domains are shown. Unique structural aspects of PKC enable it to serve as receptor for tumor promoters and antitumor-promoting agents. C1, cysteine-rich constant region in various isozymes of PKC; C1A, first 2 zinc-fingers in C1 domain; C1B, second 2 zinc-fingers in C1 domain; C2, Ca^{2+}-binding domain; C3, ATP-binding region in catalytic domain; C4, protein substrate-binding region in catalytic domain; V3, proteolysis-susceptible variable region in various isozymes; pseudosubstrate, autoinhibitory region of PKC that prevents binding of protein substrate to catalytic domain.

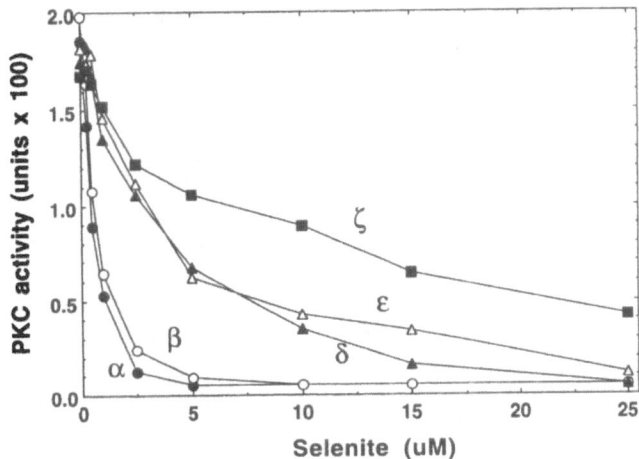

Figure 2. Differential sensitivity of various PKC isozymes to selenite. Rat brain PKC-α and -β were isolated by Ca^{2+}-dependent hydrophobic chromatography (71). PKC-δ and -ε were purified from rat brain; PKC-ζ was purified from rat kidney. Desalted PKC isozymes free from thiol agents were preincubated with indicated concentrations of selenite in wells of a multiwell plate for 5 min. Then PKC activity was determined using PKC-ε pseudosubstrate peptide.

bic association with the membrane (23). Because of the presence of cofactors in the membrane, PKC was more sensitive to selenite than in the purified form and was inactivated by low concentrations of selenite (IC_{50} = 0.05 μM). Selenite did not affect protein kinase A, phosphorylase kinase, and protein phosphatase 2A, suggesting that there is specificity in the reaction of selenite with PKC (93).

Selenoamino Acids

Selenocystine and selenodiglutathione readily inactivated the kinase activity, but not the phorbol ester binding to PKC. An inactivation of PKC in cells treated with Se-methylselenocysteine was previously reported (94). Certain selenocompounds, such as selenomethionine and methylselenocysteine, are not redox active by themselves and may require their metabolism to methylselenol (4). The redox-active selenocompounds, such as selenite, are also metabolized to methylselenol (8). This selenium metabolite has been suggested to react with cysteine residues in proteins (8). A sulfur-selenium adduct is also expected to induce a vicinal thiol oxidation.

Synthetic Selenocompounds

Synthetic organoselenium compounds, such as 1,4-phenylenebis(methylene)selenocyanate, have been reported to inhibit tumorigenesis in vivo (96) and the activities of thymidine kinase, protein kinase A, and PKC in vitro (97,98). This compound was shown to elevate GPx levels in the tissues, suggesting that selenium from this compound may be entering the assimilatory pathway (96). Ebselen (PZ51), another synthetic selenocompound with a GPx-mimetic activity, did not show any cancer-preventive activity when it was supplemented at 10 ppm in the diet (99). Previous studies showed that selenium present in ebselen was not incorporated into selenoproteins, suggesting the importance of selenium metabolism for its chemopreventive activity (100). Ebselen has also been shown to inhibit PKC activity in the test tube with an IC_{50} of 1 μM (101). However, in vitro studies have shown that ebselen has antitumor-promoting activity by inhibiting phorbol ester-induced cellular events (102). Given the fact that ebselen is poorly absorbed from the digestive tract (103), there is a possibility that the limited amount (10 ppm) of ebselen, used in the previous studies in animals, may not be sufficient to induce cancer-preventive activity (99). Nevertheless, whether ebselen, which inhibits PKC but cannot donate selenium to selenoprotein synthesis, acts as a cancer-preventive agent, if supplemented at higher doses (>10 ppm), remains to be determined.

Reversal of Selenium-Induced PKC Redox Modification by the Selenoprotein TR

The selenium-induced redox modification of PKC is reversible in different cell types to a variable extent. A protein disulfide reductase, which can reverse selenium-induced in-

activation of PKC by utilizing NADPH, is present in cells (23). Many properties of this reductase resembled TR (23). A calf thymus TR system (purified TR, thioredoxin, NADPH) regenerated the kinase activity and phorbol ester binding for PKC, which was inactivated by sulfhydryl oxidation induced by redox-active selenocompounds (104). Unexpectedly, TR induced this reduction to an appreciable extent, even in the absence of thioredoxin. Nevertheless, thioredoxin was needed for the TR-mediated regeneration of kinase activity for the proteolytically derived catalytic domain of PKC but was not required for the regeneration of the regulatory domain, having four zinc-finger thiolates. TR also regenerated the phorbol ester binding for the oxidized recombinant fragment of PKC having two zinc-fingers. Furthermore, modified TR, in which selenocysteine was either selectively alkylated or removed by carboxypeptidase treatment, was ineffective. Similarly, *Escherichia coli* TR, which is not a selenoprotein, was not effective. Conceivably, although PKC has no homology to thioredoxin, it has four zinc-fingers with the thioredoxin-type redox motif (C-X-X-C) and a flexible protein-binding region, enabling it to directly bind and react with the seleno-sulfur redox center present in TR.

PKC and Interrelationship Between Selenometabolites and Selenoproteins

It is important to consider that the selenoprotein TR can directly or indirectly reverse the inactivation of PKC induced by redox-active selenometabolites. Although the reactions of selenometabolites with PKC can induce antitumor-promoting actions, the reversal of these reactions by TR can prevent the toxic actions of selenometabolites in normal cells. In contrast, during peroxide-induced tumor promotion, because of the depletion of reducing equivalents (NADPH), TR may not function as a reductase, and thus a compromise in its action could allow the toxic action of selenometabolites, leading to the inactivation of PKC. This may provide selectivity in the action of selenometabolites to precancer cells vs. normal cells. It is also possible that in some advanced tumor cells, which are resistant to selenium, an induction of TR, along with the cellular ability to generate sufficient amounts of reducing equivalents, can give them resistance to selenium toxicity. Recent studies revealed that the PKC pathway is involved in induction of the selenoproteins TR and GPx (105–107).

Selenoproteins as a Defense for Selenium Toxicity

Previous studies have shown that selenide reacts with oxygen and induces the generation of H_2O_2 (53–55). Again, the enzyme that protects cells from this toxicity is a selenoprotein, GPx. Therefore, selenoproteins serve as a safeguard against the toxicity induced by selenometabolites (108) and also protect cells from global oxidative stress. These bidirectional effects of selenium provide a testable hypothesis to address how a selective toxicity can be achieved in precancer cells while the safety of normal tissues is maintained and how a resistance to the preventive actions of selenium is possibly developed in advanced cancer cells. Selenium actions at the low but nutritionally adequate dose are considered to be primarily mediated by selenoproteins, while toxicity to the host at higher doses of selenium is primarily mediated by selenometabolites. However, at the cancer-preventive medium, but safer, doses of selenium, its level in the body is more than that needed for the synthesis of selenoproteins. Thus, to attain selective cytotoxicity to precancer cells, there is a need for a greater retention of selenometabolites, an increase in their cytotoxic actions, or a decrease in the functions of protective selenoproteins in precancer cells compared with that in normal tissues.

PKC as a Potential Site for Selenium Interaction With Zinc and Vitamin E

Nutritional studies suggest that selenium actions are influenced by zinc and vitamin E. Zinc, by reacting with selenide, forms zinc selenide and removes excess selenium (109). Moreover, zinc, by inducing metallothionein, which can react with redox-active selenocompounds, can prevent the toxicity of selenium (110,111). Vitamin E prevents lipid peroxidation induced by diets containing high-unsaturated fat as well as that induced by selenite (108). Furthermore, selenium and vitamin E can synergistically produce anticarcinogenic actions in animals (112,113). Consistent with selenium actions that are known to be influenced by zinc and vitamin E, PKC is also regulated by these antioxidants (114,115). PKC is a zinc metalloprotein, and its activity is regulated by intracellular zinc homeostasis (114). Moreover, PKC is inhibited by vitamin E (115). These unique features, in addition to cysteine-rich regions, make PKC a more relevant target for cancer-preventive activity of selenium than other proteins that have only cysteine-rich regions.

Significance of PKC Inhibition to Cancer-Preventive Actions of Selenium

Given the facts that PKC serves as the receptor for tumor promoters and plays a crucial role in the events related to tumor promotion/progression, it may be an appropriate target for redox-active selenocompounds. The inactivation of PKC by redox modification induced by selenocompounds has functional significance in the chemopreventive actions of selenium in the various stages of carcinogenesis.

Inhibition of Tumor Promotion

Selenium has been shown to inhibit tumor promotion (20–23). Previous studies showed that an inhibition or downregulation of PKC abolished the phorbol ester-induced

induction of ornithine decarboxylase (79). Because overexpression of ornithine decarboxylase is associated with tumorigenesis (116), selenocompounds, by blocking the induction of ornithine decarboxylase and other genes via interruption of PKC function, may elicit in part their cancer-preventive action (23). Selenium-induced inhibition of PKC may also play a role in the selenium-mediated inhibition of AP-1 and NF-κB transactivation in intact cells. However, selenium is known to directly regulate AP-1 and NF-κB by oxidizing critical cysteine residues present in their DNA-binding domains (117,118). Because PKC is an important enzyme in the induction of inducible nitric oxide synthase, cyclooxygenase-2, and other enzymes involved in tumor promotion (76–80), inhibition of PKC by selenium may have a significant role in preventing the induction of these enzymes.

Inhibition of Cell Growth

Inhibition of cell proliferation is considered to be an important mechanism by which selenium inhibits carcinogenesis (56–59). Methylselenocysteine is an effective chemopreventive agent against mammary cell growth in vivo and in vitro (94). This selenocompound was shown to decrease PKC activity (94). These studies suggested PKC as an upstream target for methylselenocysteine that may trigger downstream events such as the decrease in cdk2 kinase activity and DNA synthesis, elevation of *gadd* gene expression, and finally apoptosis (94).

Induction of Apoptosis

A variety of cancer-preventive agents are believed to elicit anticarcinogenic effects, at least in part, by inducing apoptosis to remove precancerous or cancer cells that are genetically altered through mutations in oncogenes or tumor suppressor genes (119). Inhibition and/or inactivation of PKC induced by various selenocompounds may have a role in inducing apoptosis. Moreover, various commonly used PKC inhibitors, such as calphostin C, hypericin, chelerythrine, and staurosporine, induce apoptosis, which further suggests that the inactivation or inhibition of PKC triggers apoptosis (120,121).

Inhibition of PKC by its inhibitors induces apoptosis via the generation of ceramide (122,123). Furthermore, selenium-induced inactivation of PKC in prostatic carcinoma cells also leads to an elevation of ceramide and induction of apoptosis (124). PKC was shown to act as a negative modulator for sphingomyelinase and inhibit its activity (123). Therefore, an inhibition of PKC activity leads to activation of sphingomyelinase and increased generation of ceramide (123). Ceramide can increase the mitochondrial generation of reactive oxygen species and increase mitochondrial transition permeability (125). Although this can be prevented by Bcl-2, its antiapoptotic function is suppressed by a lack of phosphorylations mediated by PKC-α and mitogen-activated protein kinases (126). Then the ceramide-induced changes in mitochondria lead to the release of cytochrome c into the cytosol, where it induces the activation of caspase-3, a key protease involved in inducing apoptotic events (127). Caspase-3 activates PKC-δ by a limited proteolysis (128). Furthermore, ceramide activates PKC-ζ and c-Jun NH_2-terminal kinase, which further help in executing apoptosis. PKC isozymes, particularly PKC-α and -β, are better suited for inactivation by selenocompounds to trigger early events in apoptosis, and PKC-ζ and -δ are less susceptible for this inactivation and may facilitate the later events in apoptosis. Thus differential susceptibility of PKC isozymes to selenium is well suited for inducing apoptosis-related events.

Inhibition of Invasion and Metastasis

PKC plays an important role in regulating events related to tumor progression, such as invasion, metastasis, and tumor cell adhesion to endothelium and extracellular matrix components (129–132). Experimental tumor promoters, such as phorbol esters, and tobacco-related tumor promoters increase invasion and hematogenous metastasis (73,129, 130). Therefore, the selenium-induced inactivation of PKC may have significance in blocking these progression-related events. Selenium has been shown to decrease tumor cell invasion and attachment to matrix components in vitro (24, 133). Furthermore, selenium supplementation in the diet has been shown to decrease hematogenous metastasis (25,134).

Conclusions and Perspectives

Given the importance of dietary selenium in cancer prevention, it is important to identify the molecular targets and mechanisms by which selenium prevents cancer and exerts toxicity to the host. PKC serves as a receptor for tumor promoters, including oxidants and lipid hydroperoxides, and is activated by these agents. Redox-active selenometabolites act on the same cellular target on which the tumor promoters act, but they induce inactivation of this kinase. This may bring an efficient counteractive mechanism to block signal transduction induced by the tumor promoter at the first step. TR, a selenoprotein, can reverse this antitumor-promoting action of selenium. This suggests an interesting interrelationship between the actions of selenometabolites and selenoproteins in regulating PKC. Furthermore, PKC acts as a sensor for the induction of selenoproteins and as a site for selenium interaction with zinc and vitamin E. Selenium-induced inactivation of PKC may have significance in the cancer-preventive actions of selenium, such as inhibition of tumor promotion, cell growth, invasion, and metastasis, and in the induction of apoptosis. Therefore, in many ways, PKC is a relevant molecular target for selenium to block tumor promotion and/or early stages of tumor progression.

If PKC is acting as a target for selenium, it is important to understand why this inactivation of PKC occurs only in precancer or cancer cells, and not in normal cells, to cause toxicity to the host. It is especially important that PKC is ubiquitously distributed and plays a crucial role in many normal cellular processes. Furthermore, the concentrations of selenium that are required to achieve inhibition of various cellular processes or PKC in vitro are often higher, and only a limited concentration of selenium exists in the plasma and tissues as selenometabolites. Thus whether certain redox-cycling mechanisms can amplify the action of the selenometabolites giving specificity to precancer or cancer cells remains to be determined. Whether lipophilic and nonvolatile selenometabolites produced from some synthetic selenocompounds, such as 1,4-phenylenebis(methylene)selenocyanate, can be better retained in the membrane (98) and, thereby, can more efficiently inactivate the membrane-associated form of PKC than the natural selenocompounds, which generate the volatile metabolites, such as methylselenol or dimethylselenide, requires further study.

Acknowledgments and Notes

We thank Angela Vong and Ying Lin for assistance. Address correspondence to Rayudu Gopalakrishna, Ph.D., Dept. of Cell and Neurobiology, USC Keck School of Medicine, Los Angeles, CA 90033. Phone: (323) 442-1770. FAX: (323) 442-1771. E-mail: rgopalak@hsc.usc.edu.

Submitted 15 January 2001; accepted in final form 31 January 2001.

References

1. Clark LC: The epidemiology of selenium and cancer. *FASEB J* **44**, 2584–2589, 1985.
2. Shamberger RJ: Relationship of selenium to cancer-inhibitory effect of selenium against carcinogenesis. *JNCI* **44**, 931, 1970.
3. Combs GF Jr and Gray WP: Chemopreventive agent: selenium. *Pharmacol Ther* **79**, 179–192, 1998.
4. Ganther HE: Selenium metabolism, selenoproteins and mechanisms of cancer prevention: complexities with thioredoxin reductase. *Carcinogenesis* **20**, 1657–1666, 1999.
5. Greeder GA and Milner JA: Factors influencing the inhibitory effect of selenium on mice inoculated with Erlich ascites tumor cells. *Science* **209**, 825–826, 1980.
6. Leboeuf RA and Hoekstra WG: Changes in cellular glutathione levels: possible relation to selenium-mediated anticarcinogenesis. *Fed Proc* **44**, 2563–2567, 1985.
7. Medina D and Morrison DG: Current ideas on selenium as a chemopreventive agent. *Pathol Immunopathol Res* **7**, 187–199, 1988.
8. Ip C and Ganther HE: Activity of methylated forms of selenium in cancer prevention. *Cancer Res* **50**, 1206–1211, 1990.
9. Medina D, Lane HW, and Tracey CM: Selenium and mouse mammary tumorigenesis: an investigation of possible mechanisms. *Cancer Res Suppl* **43**, 2460s–2464s, 1983.
10. Ip C: Prophylaxis of mammary neoplasia by selenium supplementation in the initiation and promotion phases of chemical carcinogenesis. *Cancer Res* **41**, 4386–4390, 1981.
11. Pence BC, Delver E, and Dunn DM: Effects of dietary selenium on UVB-induced skin carcinogenesis and epidermal antioxidant status. *J Invest Dermatol* **102**, 759–761, 1994.
12. Perchellet JP, Abney NL, Thomas RM, Guislan YL, and Perchellet EM: Effects of combined treatment with selenium, glutathione, vitamin E on glutathione peroxidase activity, ornithine decarboxylase in-

duction and complete multistage carcinogenesis in mouse skin. *Cancer Res* **47**, 477–485, 1987.

13. Clark LC, Combs GF, Turnbull BW, Slate EH, Chalker DK, et al.: Effects of selenium supplementation for cancer prevention in patients with carcinoma of the skin. *JAMA* **276**, 1957–1963, 1996.

14. Clark LC, Dalkin B, Krongrad A, Combs GF Jr, Turnbull BW, et al.: Decreased incidence of prostate cancer with selenium supplementation: results of a double-blind cancer prevention trial. *Br J Urol* **81**, 730–734, 1998.

15. Nelson MA, Porterfiled VW, Jacobs ET, and Clark LC: Selenium and prostate cancer prevention. *Semin Urol Oncol* **17**, 91–96, 1999.

16. McCormick DL, Rao KVN, Dooley L, Steele VE, Lubet RA, et al.: Influence of *N*-methyl-*N*-nitrosourea, testosterone and *N*-(4-hydroxyphenyl)-all-*trans*-retinamide on prostate cancer induction in Wistar-Unilever rats. *Cancer Res* **58**, 3282–3288, 1998.

17. McCormick DL, Rao KVN, Steele VE, Lubet RA, Kelloff GJ, et al.: Chemoprevention of rat prostate carcinogenesis by 9-*cis*-retinoic acid. *Cancer Res* **59**, 521–524 1999.

18. Rao KVN, Johnson WD, Bosland MC, Lubet RA, Steele VE, et al.: Chemoprevention of rat prostate carcinogenesis by early and delayed administration of dehydroepiandrosterone. *Cancer Res* **59**, 3084–3089, 1999.

19. Milner JA, Pigott MA, and Dipple A: Selective effects of selenium selenite on 7,12-dimethylbenz[*a*]anthracene DNA binding in fetal mouse cell cultures. *Cancer Res* **45**, 6347–6354, 1985.

20. Borek C, Ong A, Mason H, Donahue L, and Biaglow JE: Selenium and vitamin E inhibit radiogenic and chemically induced transformation in vitro via different mechanisms of action. *Proc Natl Acad Sci USA* **83**, 1490–1494, 1986.

21. Sharma S, Stutzman JD, Kelloff GJ, and Steele VE: Screening potential chemopreventive agents using biochemical markers of carcinogenesis. *Cancer Res* **54**, 5848–5855, 1994.

22. Zhu S, Gray TE, and Nettesheim P: The effect of sodium selenite on cell proliferation and transformation of primary rat tracheal epithelial cells. *Carcinogenesis* **13**, 1725–1729, 1992.

23. Gopalakrishna R, Chen ZH, and Gundimeda U: Selenocompounds induce a redox modulation of protein kinase C in the cell, compartmentally independent from cytosolic glutathione: its role in inhibition of tumor promotion. *Arch Biochem Biophys* **348**, 37–48, 1997.

24. Gong Y and Frenkel GD: Effect of selenite on tumor cell invasiveness. *Cancer Lett* **78**, 195–199, 1994.

25. Yan L, Yee JA, Li D, McGuire MH, and Graef GL: Dietary supplementation of selenomethionine reduces metastasis of melanoma cells in mice. *Anticancer Res* **19**, 1337–1342, 1999.

26. Jiang C, Jiang W, Ip C, Ganther H, and Lu J: Selenium-induced inhibition of angiogenesis in mammary cancer at chemopreventive levels of intake. *Mol Carcinog* **26**, 213–225, 1999.

27. Medina D: Mechanisms of selenium inhibition of tumorigenesis. *J Am Coll Toxicol* **5**, 21–29, 1986.

28. Kaeck M, Lu J, Strange R, Ip C, Ganther HE, et al.: Differential induction of growth arrest inducible genes by selenium compounds. *Biochem Pharmacol* **53**, 921–926, 1997.

29. Kelloff GJ, Lieberman R, Steele VE, Boone CW, Lubet R, et al.: Chemoprevention of prostate cancer: concepts and strategies. *Eur Urol* **36**, 342–350, 1999.

30. Bosland MC: Use of animal models in defining efficacy of chemoprevention agents against prostate cancer. *Eur Urol* **36**, 459–463, 1999.

31. McCormick DL and Rao KVN: Chemoprevention of hormone-dependent prostate cancer in the Wistar-Unilever rat. *Eur Urol* **36**, 464–467, 1999.

32. Medina D and Shephard F: Selenium-mediated inhibition of mouse mammary tumorigenesis. *Cancer Lett* **8**, 241–245, 1980.

33. Ip C, Ip MM, and Kim U: Dietary selenium intake and growth of the MT-W9B transplantable rat mammary tumor. *Cancer Lett* **14**, 101–107, 1981.

34. Watrach AM, Milner JA, Watrach MA, and Poirier KA: Inhibition of human breast cancer cells by selenium. *Cancer Lett* **25**, 41–42, 1984.

35. Thompson HJ, Wilson A, Lu J, Singh M, Jiang C, et al.: Comparison of the effects of an organic and an inorganic form of selenium on a mammary carcinoma cell line. *Carcinogenesis* **15**, 183–186, 1994.

36. Lu J, Kaeck M, Jiang C, Wilson AC, and Thompson HJ: Selenite induction of DNA strand breaks and apoptosis in mouse leukemic L1210 cells. *Biochem Pharmacol* **47**, 1531–1535, 1994.

37. Ronai Z, Tillotson JK, Tranganos F, Darzynkiewicz Z, Conaway CC, et al.: Effects of organic and inorganic selenium compounds on rat mammary tumor cells. *Int Natl J Cancer* **63**, 428–434, 1995.

38. Stewart MS, Davis RL, Walsh LP, and Pence BC: Induction of differentiation and apoptosis by sodium selenite in human colonic carcinoma cells (HT29). *Cancer Lett* **117**, 35–40, 1997.

39. Wu L, Lanfear J, and Harrison PR: The selenium metabolite selenodiglutathione induces cell death by a mechanism distinct from H_2O_2 toxicity. *Carcinogenesis* **16**, 1579–1584, 1995.

40. Schwarz K, Bieri JG, Briggs GM, and Scott ML: Prevention of exudative diathesis in chicks by factor 3 and selenium. *Proc Soc Exp Biol Med* **95**, 621–625, 1957.

41. Burk RF and Hill KE: Regulation of selenoproteins. *Annu Rev Nutr* **13**, 65–81, 1993.

42. Sunde RA: Molecular biology of selenoproteins. *Annu Rev Nutr* **10**, 451–474, 1990.

43. Gladyshev VN, Jeang KT, and Stadtman TC: Selenocysteine, identified as the penultimate C-terminal residue in human T-cell thioredoxin reductase, corresponds to TGA in the human placental gene. *Proc Natl Acad Sci USA* **93**, 6146–6151, 1996.

44. Tamura T and Stadtman TC: A new selenoprotein from human lung adenocarcinoma cells: purification, properties and thioredoxin reductase activity. *Proc Natl Acad Sci USA* **93**, 1006–1011, 1996.

45. Liu SY and Stadtman TC: Heparin-binding properties of selenium-containing thioredoxin reductase from HeLa cells and human lung adenocarcinoma cells. *Proc Natl Acad Sci USA* **94**, 6138–6141, 1997.

46. Berggren MM, Mangin JF, Gasdaska JR, and Powis G: Effect of selenium on rat thioredoxin reductase actvity. *Biochem Pharmacol* **57**, 187–193, 1997.

47. Hill KE, McCollum GW, Boeglin ME, and Burk RF: Thioredoxin reductase activity is decreased by selenium deficiency. *Biochem Biophys Res Commun* **234**, 293–295, 1997.

48. Gallegos A, Berggren M, Gasdasaka JR, and Powis G: Mechanisms of the regulation of thioredoxin reductase activity in cancer cells by the chemopreventive agent selenium. *Cancer Res* **57**, 4965–4970, 1997.

49. Spyrou G, Bjornstedt M, Skog S, and Holmgren A: Selenite and selenate inhibit human lymphocyte growth via different mechanisms. *Cancer Res* **56**, 4407–4412, 1996.

50. Baker A, Payne CM, Briehl MM, and Powis G: Thioredoxin, a gene found overexpressed in human cancer, inhibits apoptosis in vitro and in vivo. *Cancer Res* **57**, 5162–5167, 1997.

51. Hoffman ER, Boyanapalli M, Lindner DJ, Wei X, Hassel BA, et al.: Thioredoxin reductase mediates cell death effects of the combination of β-interferon and retinoic acid. *Mol Cell Biol* **18**, 6493–6504, 1998.

52. Heppell-Parton AA, Cahn A, Bench A, Lowe N, Lehrach H, et al.: Thioredoxin, a mediator of growth inhibition maps to 9q31. *Genomics* **26**, 379–381, 1995.

53. Spallholz JE: On the nature of selenium toxicity and carcinostatic activity. *Free Radic Biol Med* **17**, 45–64, 1994.

54. Garberg P, Stehl A, Warholm M, and Hogberg J: Studies of the role of DNA fragmentation in selenium toxicity. *Biochem Pharmacol* **37**, 3401–3406, 1988.

55. LeBoeuf RA and Hoekstra WG: Adaptive changes in hepatic glutathione metabolism in response to excess selenium in rats. *J Nutr* **113**, 845–854, 1983.

56. Medina D and Oborn CJ: Selenium inhibition of DNA synthesis in mouse mammary epithelial cell line YN-4. *Cancer Res* **44**, 4361–4365, 1984.

57. Fico ME, Poirier KA, Watrach AM, and Milner JA: Differential effects of selenium on normal and neoplastic canine mammary cells. *Cancer Res* **46**, 3384–3388, 1986.

58. Medina D and Oborn CJ: Differential effects of selenium on the growth of mouse mammary cells in vitro. *Cancer Lett* **13**, 333–344, 1981.

59. Frenkel GD and Falvey D: Evidence for the involvement of sulfhydryl compounds in the inhibition of cellular DNA synthesis by selenite. *Mol Pharmacol* **34**, 573–577, 1988.

60. Whanger P, Vendeland S, Park YC, and Xia Y: Metabolism of subtoxic levels of selenium in animals and humans. *Ann Clin Lab Sci* **26**, 99–113, 1996.

61. Levander OA: Considerations in the design of selenium bioavailability studies. *Fed Proc* **42**, 1721–1725, 1983.

62. Sandholm M: Function of erythrocytes in attaching selenite-Se onto specific plasma proteins. *Acta Pharmacol Toxicol* **36**, 321–327, 1975.

63. Redman C, Scot JA, Bainec AT, Basye JL, Clark LC, et al.: Inhibitory effect of selenomethionine on the growth of three selected human tumor cell lines. *Cancer Lett* **125**, 103–110, 1998.

64. Redman C, Xu MJ, Peng YM, Scot JA, Payne C, et al.: Involvement of polyamines in selenomethionine-induced apoptosis and mitotic alteration in human tumor cells. *Carcinogenesis* **18**, 1195–1202, 1997.

65. Waschulewski IH and Sunde RA: Effect of dietary methionine on utilization of tissue selenium from dietary selenomethionine from glutathione peroxidase in the rat. *J Nutr* **118**, 367–374, 1987.

66. Deagan J, Butler J, Zachara B, and Whanger P: Determination of the distribution of selenium between glutathione peroxidase, selenoprotein P, and albumin in plasma. *Anal Biochem* **208**, 176–181, 1993.

67. Deagan J, Butler J, Zachara B, and Whanger P: Determination of the distribution of selenium between glutathione peroxidase, selenoprotein P, and albumin in plasma. *Anal Biochem* **208**, 176–181, 1993.

68. Nishizuka Y: Intracellular signaling by hydrolysis of phospholipids and activation of protein kinase C. *Science* **258**, 607–614, 1992.

69. Jaken S: Protein kinase C and tumor promoters. *Curr Opin Cell Biol* **2**, 192–197, 1990.

70. Blumberg PM: Complexities of the protein kinase C pathway. *Mol Carcinog* **4**, 330–344, 1991.

71. Gopalakrishna R and Anderson WB: Ca^{2+}- and phospholipid-independent activation of protein kinase C by selective oxidative modification of regulatory domain. *Proc Natl Acad Sci USA* **86**, 6758–6762, 1989.

72. Gopalakrishna R and Anderson WB: Reversible oxidative activation and inactivation of protein kinase C by the mitogen/tumor promoter periodate. *Arch Biochem Biophys* **285**, 382–387, 1991.

73. Gopalakrishna R, Chen ZH, and Gundimeda U: Tobacco smoke tumor promoters, catechol and hydroquinone, induce oxidative regulation of protein kinase C and influence invasion and metastasis of lung carcinoma cells. *Proc Natl Acad Sci USA* **91**, 12233–12237, 1994.

74. Kass GEN, Duddy SK, and Orrenius S: Activation of hepatocyte protein kinase C by redox-cycling quinones. *Biochem J* **260**, 499–507, 1989.

75. Larsson R and Cerutti P: Translocation and enhancement of phosphotransferase activity of protein kinase C following exposure of mouse epidermal cells to oxidants. *Cancer Res* **49**, 5627–5633, 1989.

76. Meyer M, Schreck R, Muller JM, and Baeuerle PA: Redox control of gene expression by eukaryotic transcription factors NF-κB, AP-1 and SRF/TCF. In *Oxidative Stress, Cell Activation and Viral Infection*, Pasquier C, Auclair C, Olivier RY, and Packer L (eds). Switzerland: Birkhauser Verlag, 1994, pp 217–235.

77. Stauble B, Boscoboinik D, Tasinato A, and Azzi A: Modulation of activator protein-1 (AP-1) transcription factor and protein kinase C by hydrogen peroxide and α-tocopherol in vascular smooth muscle cells. *Eur J Biochem* **226**, 393–402, 1994.

78. Amstad PA, Krupitza G, and Cerutti PA: Mechanism of c-*fos* induction by active oxygen. *Cancer Res* **52**, 3952–3960, 1992.

79. Fischer SM, Lee ML, Maldve RE, Morris RJ, Trono D, et al.: Association of protein kinase C activation with induction of ornithine decarboxylase in murine but not human keratinocyte cultures. *Mol Carcinog* **7**, 228–237, 1993.

80. Klein IK, Ritland SR, Burgart LJ, Ziesmer SC, Roche PC, et al.: Adenoma-specific alterations of protein kinse C isozyme expression in APC^min mice. *Cancer Res* **60**, 2077–2080, 2000.

81. Gopalakrishna R and Jaken S: Protein kinase C signaling and oxidative stress. *Free Radic Biol Med* **28**, 1349–1361, 2000.

82. Gopalakrishna R, Chen ZH, and Gundimeda U: Modification of cysteine-rich regions in protein kinase C induced by oxidant tumor promoters and the enzyme-specific inhibitors. *Methods Enzymol* **252**, 134–148, 1995.

83. Quest AFG, Bloomenthal J, Bardes ESG, and Bell RM: The regulatory domain of protein kinase C coordinates four atoms of zinc. *J Biol Chem* **267**, 10193–10197, 1992.

84. Kazanietz MG, Wang S, Milner GWA, Lewin NE, Liu HL, et al.: Residues in the second cysteine-rich region of protein kinase C relevant to phorbol ester binding as revealed by site-directed mutagenesis. *J Biol Chem* **270**, 21852–21859, 1995.

85. O'Brian CA, Ward NE, Weinstein IB, Bull AW, and Marnett LJ: Activation of rat brain protein kinase C by lipid oxidation products. *Biochem Biophys Res Commun* **155**, 1374–1380, 1988.

86. Liu B, Timar J, Howlett J, Diglio CA, and Honn KV: Lipoxygenase metabolites of arachidonate and linoleic acids modulate the adhesion of tumor cells to endothelium via regulation of protein kinase C. *Cell Regul* **2**, 1045–1055, 1991.

87. Sweetman LL, Zhang NY, Peterson H, Gopalakrishna R, and Sevanian A: Effect of linoleic acid hydroperoxide on endothelial cell calcium homeostasis and phospholipid hydrolysis. *Arch Biochem Biophys* **323**, 97–107, 1995.

88. Gopalakrishna R, Chen Z, Gundimeda U, and Kensler T: Reversible inactivation of protein kinase C by oltipraz through reaction with the catalytic domain: role in inhibition of tumor promotion (abstr). *Proc Am Assoc Cancer Res* **37**, A1817, 1996.

89. Gundimeda U, Chen Z, and Gopalakrishna R: Tamoxifen modulates protein kinase C via oxidative stress in estrogen receptor-negative breast cancer cells. *J Biol Chem* **271**, 13504–13514, 1996.

90. Chen Z, Gundimeda U, and Gopalakrishna R: Curcumin irreversibly inactivates protein kinase C activity and phorbol ester binding: its possible role in cancer chemoprevention (abstr). *Proc Am Assoc Cancer Res* **37**, 282, 1996.

91. Boscoboinik D, Szewczyk A, Hensey C, and Azzi A: Inhibition of cell proliferation by α-tocopherol: role of protein kinase C. *J Biol Chem* **266**, 6188–6194, 1991.

92. Gopalakrishna R, Gundimeda U, and Chen Z: Vitamin E succinate inhibits protein kinase C: correlation with its unique inhibitory effects on cell growth and transformation. In *Nutrients and Cancer*, Prasad KN et al. (eds). Totowa, NJ: Humana, 1995, pp 21–37.

93. Gopalakrishna R, Gundimeda U, and Chen Z: Cancer-preventive selenocompounds induce a specific redox modification of cysteine-rich regions in Ca^{2+}-dependent isoenzymes of protein kinase C. *Arch Biochem Biophys* **348**, 25–36, 1997.

94. Sinha R, Kiley SC, Lu JX, Thompson JJ, Moraes R, et al.: Effects of methylselenocysteine on PKC activity, cdk2 phosphorylation and *gadd* gene expression in synchronized mouse mammary epithelial tumor cells. *Cancer Lett* **146**, 135–145, 1999.

95. Ganther HE and Corcoran C: Selenotrisulfides. II. Cross-linking of reduced pancreatic ribonuclease with selenium. *Biochemistry* **8**, 2557–2563, 1969.

96. El-Bayoumy K, Upadhyaya P, Chae YH, Sohn OS, Rao CV, et al.: Chemoprevention of cancer by organoselenium compounds. *J Cell Biochem Suppl* **22**, 92–100, 1995.

97. Tillotson JK, Upadhyaya P, and Ronai Z: Inhibition of thymidine kinase in cultured mammary cells by the chemopreventive organoselenium compound, 1,4-phenylenebis(methylene)selenocyanate. *Carcinogenesis* **15**, 607–610, 1994.

98. Foiles PG, Fujiki H, Suganuma M, Okabe S, Yatsunami J, et al.: Inhibition of PKC and PKA by chemopreventive organoselenium compounds. *Int J Oncol* **7**, 685–690, 1995.

99. Ip C and Ganther HE: Activity of methylated forms of selenium in cancer prevention. *Cancer Res* **50**, 1206–1211, 1990.

100. Wendel A, Fausel M, Safayhi H, Tiegs G, and Otter RA: Novel biologically active seleno-organic comound. II. Activity of PZ51 in relation to glutathione peroxidase. *Biochem Pharmacol* **33**, 3241–3245, 1984.

101. Cotgreave IA, Duddy SK, Kass GEN, Thompson D, and Moldeus P: Studies on the anti-inflammatory activity of ebselen: ebselen intereferes with granulocyte oxidative burst by dual inhibition of NADPH oxidase and protein kinase C. *Biochem Pharmacol* **38**, 649–656, 1989.

102. Hu J, Engman L, and Cotgreave IA: Redox-active chalcogen-containing glutathione peroxidase mimetics and antioxidants inhibit tumor promoter-induced downregulation of gap junctional intercellular communication between WBF344 liver epithelial cells. *Carcinogenesis* **16**, 1815–1824, 1995.

103. Sies H and Masumoto H: Ebselen as a glutathione peroxidase mimic and as a scavenger of peroxynitrite. *Adv Pharmacol* **38**, 229–246, 1997.

104. Gopalakrishna R and Gundimeda U: Selenoprotein thioredoxin reductase activates the oxidized protein kinase C by a direct interaction with its zinc-thiolates (abstr). *FASEB J* **14**, A296, 2000.

105. Jornot I and Junod AF: Hyperoxia, unlike phorbol ester, induces glutathione peroxidase through a protein kinase C-independent mechanism. *Biochem J* **326**, 117–123, 1997.

106. Kumar S and Holmgren A: Induction of thioredoxin, thioredoxin reductase and glutaredoxin activity in mouse skin by TPA, a calcium ionophore and other tumor promoters. *Carcinogenesis* **20**, 1761–1767, 1999.

107. Anema SM, Walker SW, Howie AF, Arthur JR, Nicol F, et al.: Thioredoxin reductase is the major selenoprotein expressed in human umbilical vein endothelial cells and is regulated by protein kinase C. *Biochem J* **342**, 111–117, 1999.

108. Dougherty JJ and Hoekstra WG: Stimulation of lipid peroxidation in vivo by injecting selenite and lack of stimulation by selenite. *Proc Soc Exp Biol Med* **169**, 209–215, 1982.

109. House WA and Welch RM: Bioavailability of and interaction between zinc and selenium in rats fed wheat grain intrinsically labeled with ^{65}Zn and ^{75}Se. *J Nutr* **119**, 916–921, 1989.

110. Liu J, Kershaw WC, and Klaassen CD: Protective effects of zinc on cultured rat primary hepatocytes to metals with low affinity for metallothionein. *J Toxicol Environ Health* **35**, 51–62, 1992.

111. Lazo JS, Kondo Y, Dellapiazza D, Michalska AE, Choo KHA, et al.: Enhanced sensitivity of oxidative stress in cultured embryonic cells from transgenic mice deficient in metallothionein I and II genes. *J Biol Chem* **270**, 5506–5510, 1995.

112. Horvath PM and Ip C: Synergistic effect of vitamin E and selenium in the chemoprevention of mammary carcinogenesis in rats. *Cancer Res* **43**, 5335–5341, 1983.

113. Ip C and White G: Mammary cancer chemoprevention by inorganic and organic selenium: single agent treatment or in combination with vitamin E and their effects on in vitro immune functions. *Carcinogenesis* **8**, 1763–1766, 1987.

114. Csermely P, Szamel M, Resch K, and Somogyi J: Zinc can increase the activity of protein kinase C and contributes to its binding to plasma membranes in T lymphocytes. *J Biol Chem* **263**, 6487–6490, 1988.

115. Boscoboinik D, Szewczyk A, Hensey C, and Azzi A: Inhibition of cell proliferation by α-tocopherol: role of protein kinase C. *J Biol Chem* **266**, 6188–6194, 1991.

116. O'Brien TG, Megosh LC, Gilliard G, and Soler AP: Ornithine decarboxylase overexpression is a sufficient condition for tumor promotion in mouse skin. *Cancer Res* **57**, 2630–2637, 1997.

117. Spyrou G, Bjornstedt M, Kumar S, and Holmgren A: AP-1 DNA-binding activity is inhibited by selenite and selenodiglutathione. *FEBS Lett* **368**, 59–63, 1995.

118. Kim IY and Stadtman TC: Inhibition of NF-κB DNA binding and nitric oxide induction in human T cells and lung carcinoma cells by selenite treatment. *Proc Natl Acad Sci USA* **94**, 12904–12907, 1997.

119. Kelloff GJ, Boone CW, Steele VE, Crowell JA, Greenwald P, et al.: *Mechanistic Considerations in the Evaluations of Chemopreventive Data*. Lyon, France: Int Agency Res Cancer, 1996, pp 203–219. (IARC Publ 139)

120. Jarvis WD, Turner AJ, Povirk LF, Taylor RS, and Grant S: Induction of apoptotic DNA fragmentation and cell death in HL-60 human promyelocytic leukemia cells by pharmacological inhibitors of protein kinase C. *Cancer Res* **54**, 1707–1714, 1994.

121. Whelan RD and Parker PJ: Loss of protein kinase C function induces a proapoptotic response. *Oncogene* **16**, 1939–1944, 1998.

122. Hannun YA: Functions of ceramide in coordinating cellular responses to stress. *Science* **274**, 1855–1859, 1996.

123. Haimovitz-Friedman A, Kolesnick RN, and Fuks Z: Ceramide signaling in apoptosis. *Br Med Bull* **53**, 539–553, 1997.

124. Gopalakrishna R and Gundimeda U: Selenium-induced apoptosis in prostate cancer cells: role of protein kinase C inactivation and ceramide generation (abstr). *Proc Am Assoc Cancer Res* **41**, 341, 2000.

125. Bradham C, Quian T, Streetz K, Trautwein C, Brenner DA, et al.: The mitochondrial permeability transition is required for tumor necrosis factor α-mediated apoptosis and cytochrome c release. *Mol Cell Biol* **18**, 6353–6364, 1998.

126. Ruvolo PP, Deng X, Carr BK, and May WSA: Functional role for mitochondrial protein kinase C in Bcl2 phosphorylation and suppression of apoptosis. *J Biol Chem* **273**, 25436–25442, 1998.

127. Kluck RM, Bossy-Wetzel E, Green DR, and Newmeyer DD: The release of cytochrome c from mitochondria: a primary site of Bcl-2 regulation of apoptosis. *Science* **275**, 1132–1136, 1997.

128. Denning MF, Wang Y, Nickoloff BJ, and Wrone-Smith T: Protein kinase Cδ is activated by caspase-dependent proteolysis during ultraviolet radiation-induced apoptosis of human keratinocytes. *J Biol Chem* **273**, 29995–30002, 1998.

129. Gopalakrishna R and Barsky SH: Tumor promoter-induced membrane-bound protein kinase C regulates hematogenous metastasis. *Proc Natl Acad Sci USA* **85**, 612–616, 1988.

130. Korczak B, Whale C, and Kerbel RS: Possible involvement of Ca^{2+} mobilization and protein kinase C activation in the induction of spontaneous metastasis by mouse mammary adenocarcinoma. *Cancer Res* **49**, 2597–2560, 1989.

131. Liu B and Honn KV: Protein kinase C inhibitor calphostin C inhibits B16 melanoma metastasis. *Int J Cancer Res* **52**, 147–153, 1992.

132. Kiley SC, Clark KJ, Goodnough M, Welch DR, and Jaken S: Protein kinase C δ involvement in mammary tumor cell metastasis. *Cancer Res* **59**, 3230–3238, 1999.

133. Yan L and Frenkel GD: Inhibition of cell attachment by selenite. *Cancer Res* **52**, 5803–5807, 1992.

134. Tanaka T, Kohno H, Murakami M, Kagami S, and El-Bayoumy K: Suppressing effects of dietary supplementation of the organoselenium 1,4-phenylenebis(methylene)selenocyanate and the citrus antioxidant auraptene on lung metastasis of melanoma cells in mice. *Cancer Res* **60**, 3713–3716, 2000.

NUTRITION AND CANCER, *40*(1), 64–73

Antiangiogenic Activity of Selenium in Cancer Chemoprevention: Metabolite-Specific Effects

Junxuan Lu and Cheng Jiang

Abstract: We review recent data that support a potential antiangiogenic effect of selenium (Se) in the chemoprevention of cancer and data that contrast two pools of Se metabolites, namely, methylselenol vs. hydrogen selenide, that differentially affect proteins and cellular processes crucial to tumor angiogenesis regulation. With regard to tumor angiogenesis, the chemopreventive effect of increased Se intake on chemically induced mammary carcinogenesis has been associated with reduced intratumoral microvessel density and an inhibition of the expression of vascular endothelial growth factor. The in vitro data show that monomethyl Se potently inhibits cell cycle progression of vascular endothelial cells to the S phase, endothelial expression of matrix metalloproteinase-2, and cancer epithelial expression of vascular endothelial growth factor with concentrations giving half-maximal inhibition that are within the plasma range of Se in US adults. The methyl Se-specific activities may therefore be physiologically pertinent for angiogenic switch regulation in early lesions in vivo in the context of cancer chemoprevention, which aims at retarding and blocking the growth and progression of early lesions. We argue for the antiangiogenic action of Se, especially the methyl Se pool of metabolites, as a primary mechanism for preventing avascular lesion growth. Contrary to the currently held paradigm, we speculate that there is a potential role for selenoproteins in regulating the growth and fate of transformed epithelial cells.

Introduction

The micronutrient selenium (Se) has long been implicated to have an anticancer potential by epidemiological and laboratory studies. The landmark cancer prevention trial by Larry Clark, Gerald Combs, Jr., and co-workers demonstrated for the first time that an Se supplement (200 µg/day) provided as selenized yeast (Se-yeast) to a skin cancer patient population otherwise adequate in Se nutrition might be a safe and effective preventive agent for several major human epithelial cancers, including those of the prostate, lung, and colon, while lacking protective activity against second primary skin cancer (Ref. 1 and updates in this issue). A community-scale intervention study and a small-scale clini-

cal trial in China also indicated preventive efficacy of Se against liver cancer (2). Encouraged by these results, two new clinical trials sponsored by the National Cancer Institute have begun recruiting patients to verify the preventive efficacy of Se-yeast or its major constituent selenomethionine for prostate cancer (Southwest Oncology Group, SWOG Protocol S0000) and lung cancer (Eastern Cooperative Oncology Group, ECOG Protocol E5597). A better understanding of the mechanisms through which Se exerts anticancer activity will be important for helping to interpret the outcomes of these new "definitive" trials and to develop safer and more effective Se agents for cancer prevention, especially in organ site-specific manners. At this crucial juncture of Se translation research, this special issue as a fitting tribute to Larry Clark for his pioneering work provides a timely review and synthesis of the status of the field and a forum to discuss and define future directions.

On a personal note, one of us ("Johnny" Lu) became acquainted with Larry Clark in 1984 at Cornell University as a fresh graduate student of Jerry Combs. Johnny even helped translate some materials from Chinese scientific literature when data were being gathered to support the landmark Se trial (1) as we know it today. While working on zinc toxicity for his Ph.D. degree, Johnny witnessed a lot of human plasma samples being analyzed for Se in Jerry's laboratory for that trial although, at the time, without much appreciation of the significance of the undertaking. That changed significantly after he embarked on a cancer research career. He holds great respect and admiration for the leadership and perseverance of Larry and his collaborators for bringing that trial to fruition and for the determination and courage displayed by Larry for continuing to push for additional Se trials even when he was dying of prostate cancer. In honor of Larry's memory, we review our recent work on Se and angiogenesis regulation as a potential novel mechanism of cancer prevention by Se.

Known "Mechanisms" of Se Anticancer Activity

Even though, for convenience reasons, the anticancer potential of Se is often described in terms of the element, a vast

The authors are affiliated with the AMC Cancer Research Center, Denver, CO 80214.

literature base indicates that it is expressed as a function of the dose and chemical form in which the element resides, not elemental Se per se (3,4). With respect to "mechanisms," a number of them have been investigated: antioxidant protection (via SeCys-glutathione peroxidases), altered carcinogen metabolism, enhanced immune surveillance, cell cycle effects, enhanced apoptosis (3,4), and, more recently, inhibition of neoangiogenesis (5,6). The mechanisms that are actually involved in cancer prevention by Se will likely depend on the dose and form of Se compounds, the Se status of the individual, perhaps the type and etiology of malignancy, and even the organ sites. It is probable that Se supplementation of individuals with relatively low or frankly deficient Se intakes can be expected to support enhanced antioxidant protection as a result of increased expression of the SeCys enzymes or enhanced immune surveillance (4).

On the other hand, in the trial of Clark et al. (1) and in animal models (3,4), cancer-preventive effects have been observed at Se intakes that are more than sufficient to correct nutritional deficiency. That is, Se appears to be antitumorigenic at intake levels that are substantially higher than those associated with maximal expression of the known SeCys-containing glutathione peroxidase enzymes (3,4). In this context of cancer chemoprevention, methylselenol or related monomethyl Se species have been implicated as a candidate in vivo active metabolite pool (3,7,8). We and others have used cell culture models to seek a better understanding of how the different pools of Se inhibit survival of tumor cells through apoptosis induction (9–12). This aspect of the work has been reviewed recently (13). [For additional information related to apoptosis induction, see pertinent articles in this issue and our recent work documenting caspase-dependent apoptosis execution pathways induced by the methyl Se pool (14).]

We focus on the regulation of angiogenesis by Se, especially by specific Se metabolite pools in cancer prevention. We limit the scope of discussion to those forms of Se that feed into the genotoxic hydrogen selenide pool or the nongenotoxic methylselenol pool of Se metabolites (Fig. 1) (6,13). There is no literature documentation on the antiangiogenic attributes of other Se forms, such as synthetic aromatic organo-Se compounds. We review pertinent data that support methylselenol-specific inhibitory activities on angiogenic switch mechanisms in vascular endothelial cells [endothelial mitogenesis and matrix metalloproteinase (MMP) expression] and in the tumor epithelial cells (angiogenic cytokine expression) (Fig. 1). We discuss a model of cancer prevention by Se based on the interaction of epithelial lesions and the vasculature that supports such lesions.

Angiogenesis Is Obligatory for Carcinogenesis

Carcinogenesis is a multistep process of tumor initiation, promotion, progression, and metastasis. Human cancers likely involve a gradual accumulation of genetic and epigenetic changes over a period of decades. Because cancer

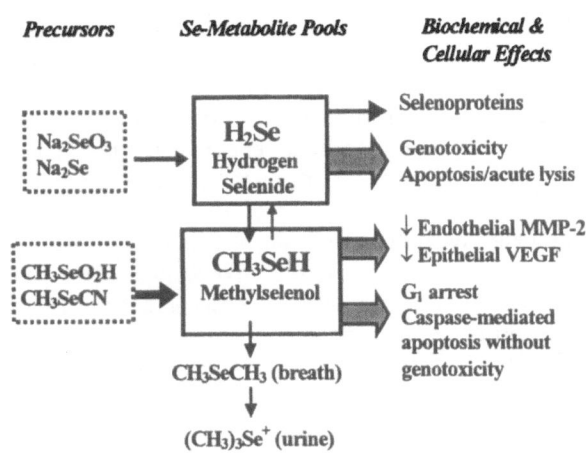

Figure 1. Schematic relationship of Se precursors feeding into genotoxic hydrogen selenide pool or nongenotoxic methylselenol pool of metabolites that exert distinct biochemical and cellular effects. Inhibitory activities by methylselenol pool on protein molecules and cellular processes relevant to angiogenesis support Se metabolite-specific mechanisms for angiogenic switch control (6). Genotoxicity (or lack of) effect of the 2 Se pools was based on studies with mammary cancer epithelial cells and leukemia cells (13). CH$_3$SeO$_2$H, methylseleninic acid; CH$_3$SeCN, methylselenocyanate; MMP-2, matrix metalloproteinase-2; VEGF, vascular endothelial growth factor. (Modified from Refs. 6 and 13.)

initiation is arguably inevitable, the major thrust of prevention research in animal tumor models and in clinical cancer trials has been on targeting promotion and progression stages. Angiogenesis, i.e., the process of formation of new microvessels from existing vessels, is a critical and obligatory component of promotion, progression, and metastasis of solid cancers, most of which are of epithelial origin (15–18). At the onset of solid tumor genesis, initiated cells undergo clonal expansion in an avascular state, because the expanding lesions are small enough to take in nutrients and to expel metabolic wastes by diffusion. However, diffusion is not sufficient to support any continued growth of the lesion beyond a certain physical size (estimated to be a few mm^3), because the expanding lesions consume nutrients at a rate proportional to their volume, whereas the supply of nutrients is delivered at a rate proportional to their surface area. It has been estimated that such "dormant" avascular lesions exist in high prevalence in organs of asymptomic adults. In order for lesions to grow beyond this size limit imposed by simple diffusion, they must turn on their angiogenic switch (15–17) to form neovasculature to provide the necessary nutrients and factors and remove metabolite wastes. Therefore, tumor angiogenic switch control can provide a logical and attractive target for the chemoprevention of cancer.

Angiogenesis Is Regulated Through Epithelial and Endothelial Compartments

Angiogenesis in solid epithelial tumors is controlled through at least two principal cell compartments: the transformed epithelial cells, which serve as a main source of angiogenic factors, and the vascular endothelial cells, which constitute the targets for the angiogenic signals (15–18). On the tumor cell side, the production of angiogenic stimulatory cytokines by cancer epithelial cells is a major means of controlling the angiogenic switch. Most prominent among the positive factors is vascular endothelial growth factor (VEGF)/vascular permeability factor (19,20). Cancer epithelial cells are the main source of VEGF expression (21–25), although recent data indicate that other cell types can also express it in a tumorigenic environment (26). VEGF plays a crucial role in vasculogenesis and angiogenesis, since in germ-line knockout experiments, a loss of even one VEGF allele leads to the abnormal formation of intra- and extraembryonic vessels and embryonic lethality, indicating a tight dose-dependent regulation of embryonic vessel development by VEGF (27,28). Furthermore, VEGF-null embryonic stem cells exhibit a dramatically reduced ability to form tumors in nude mice (28). Whereas overexpression of VEGF is linked to increased angiogenesis and more aggressive tumor behavior (29,30), antiangiogenic interventions, especially those based on VEGF antibodies or interference of signal transduction through its receptors (31–35), have been shown to result in the inhibition of tumor growth and induction of endothelial apoptosis. Because hypoxia is common in avascular lesions (36) and because hypoxia (37) and certain oncogenic mutations (38–41) and autocrine or paracrine growth factors in the tumor (42–44) are potent inducers of its expression, VEGF is a probable primary stimulator for the angiogenic switch in early lesions, which are the likely responsive targets of chemoprevention.

On the vascular side, the endothelial cells respond to angiogenic stimulation through three key processes (15–17): 1) increased expression and secretion of MMPs (45) to break down the outer sheath and extracellular matrix of existing vessels for the endothelial cells to sprout through, 2) increased cell motility for remodeling and invasion during the sprouting process, and 3) mitogenic signaling leading to cell cycle entry and cell division to provide the number of cells necessary for elongating the new sprout. The crucial role of specific MMPs, such as MMP-2, in tumor angiogenesis has been documented using knockout as well as other model approaches (46,47). Inhibiting one or more of these processes can negatively impact the endothelial response(s) to a given angiogenic signal. As discussed below, methylselenol precursors inhibit with significant potency the endothelial expression of MMP-2, its mitogenesis stimulated by angiogenic factors, and its survival via caspase-mediated apoptosis.

In Vivo Findings

Antiangiogenic Activity Was Associated With Mammary Cancer Prevention by Se

We recently initiated work to explore the hypothesis that Se may exert cancer-chemopreventive activity, at least in part, through an antiangiogenic mechanism (5). In a chemoprevention setting, Se (3 ppm) as Se-garlic (*experiment 1*) or selenite (*experiment 2*) was fed for ~2 mo to Sprague-Dawley rats that were given a single intraperitoneal injection of methylnitrosourea to initiate mammary carcinogenesis 1 wk earlier. The microvessels in the mammary tumors were visualized with immunohistochemical staining for factor VIII, and the microvessel number (counts/0.5 mm², 10 fields) in "hot-spot" stromal areas was counted. Mammary carcinomas in the Se-fed rats was 34% (*experiment 1*) and 24% (*experiment 2*) lower than in rats fed the control diet (Table 1). When categorized by the size of the microvessels, the reduction of microvessel density in the Se-fed groups was almost exclusively confined to the small microvessels (1–4 cells in diameter). The microvessel density of the uninvolved mammary glands was not decreased by Se-garlic treatment (Table 1). Similar results were obtained when established

Table 1. Effects of a Chemopreventive Level of Dietary Se as Se-Garlic or Selenite on Microvessel Density of NMU-Induced Rat Mammary Carcinomas and Noninvolved Mammary Glands[a,b]

	n	Density, counts/0.5 mm²			
		Large vessels	Medium vessels	Small vessels	Total vessels
Experiment 1					
Carcinomas					
Control	9	5 ± 1	10 ± 1	55 ± 6	69 ± 6
Se-garlic	6	3 ± 1	8 ± 2	35 ± 6*	46 ± 6*
Mammary glands					
Control	6	1.8 ± 0.5	2.7 ± 0.4	4.2 ± 0.8	8.7 ± 0.7
Se-garlic	6	1.3 ± 0.4	2.0 ± 0.7	3.8 ± 0.6	7.2 ± 0.9
Experiment 2					
Carcinomas					
Control	8	0.9 ± 0.4	4 ± 2	75 ± 5	80 ± 4
Selenite	4	0.3 ± 0.3	4 ± 3	57 ± 2*	61 ± 3*

a: Values are means ± SE. Vessels are classified as follows: >9 cells in diameter (large), 5–9 cells in diameter (medium), and 1–4 cells in diameter (small). NMU, 1-methyl-1-nitrosourea. Data are from Ref. 5.

b: Statistical significance is as follows: *, significantly different from control, $P < 0.05$.

mammary carcinomas were treated acutely through bolus doses of Se (5). These results indicated a potential anti-angiogenic effect of chemopreventive intake of Se and that the effect was neoplasia specific. Because growing and newly sprouted microvessels are likely to be smaller, the observed reduction of small vessels by Se treatments indicated that a mechanism(s) governing the genesis of new vessels might be inhibited.

Se Decreased Expression of VEGF in Some Carcinomas

The expression level of VEGF in mammary carcinomas was measured by Western blot analyses (5). On the basis of the limited number of samples analyzed, two of five carcinomas in the Se-garlic group and two of four carcinomas in the selenite group showed a marked reduction in VEGF expression to almost nondetectable levels. These results indicated that VEGF downregulation might be involved in some, but not all, tumors. Similar to the chemoprevention setting, acute Se treatment of established mammary carcinomas showed a marked reduction of VEGF expression in some, but not all, treated carcinomas (5).

In Vitro Metabolite-Specific Antiangiogenic Attributes

To define the mechanisms underlying the antiangiogenic activity of chemopreventive levels of Se intake, we have examined the effects of direct Se exposure in cell culture on the expression of VEGF by cancer epithelial cells, the expression of MMP-2 in human umbilical vein endothelial cells (HUVECs) (6), and the mitogenesis and survival of HUVECs (48). A number of methylselenol precursors, including methylseleninic acid (MSeA) and methylselenocyanate (MSeCN), were used to enrich the methylselenium (methyl Se) pool in vitro. The results suggest a methyl-selenol-specific inhibition of the angiogenic switch mechanisms through multiple processes.

Methyl Se-Specific Inhibitory Effect on VEGF Expression

We recently showed that brief exposure of human DU 145 prostate cancer cells to increasing concentrations of MSeA decreased both cellular (Fig. 2A) and secreted VEGF levels in an Se concentration-dependent manner (6), with the concentration resulting in half-maximal inhibition (IC_{50}) of ~2 μM, which is within range of plasma Se concentration in most US residents (mean ~1.5 μM) (1). The inhibitory effect of MSeA on VEGF expression was independent of serum level in the medium and the confluence status of the cells and was detectable within 2 h of exposure (Fig. 2B). In contrast, exposure to selenite in the same dose range or higher did not decrease VEGF expression.

Longer exposure of DU 145 cells to MSeA or sodium selenite above some threshold levels led to apoptosis as indicated by DNA nucleosomal fragmentation (Fig. 2C). The inhibitory effect of MSeA on VEGF expression was ob-

A. DU-145 cellular VEGF, 6 h

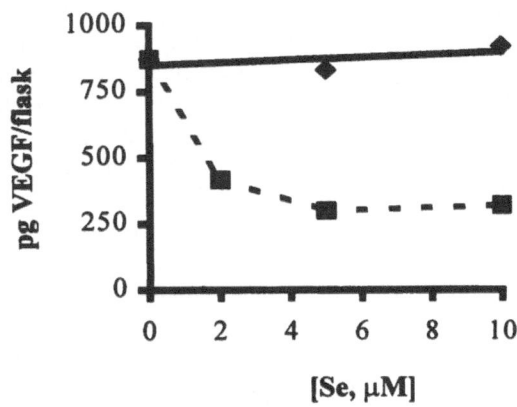

B. Time course, secreted

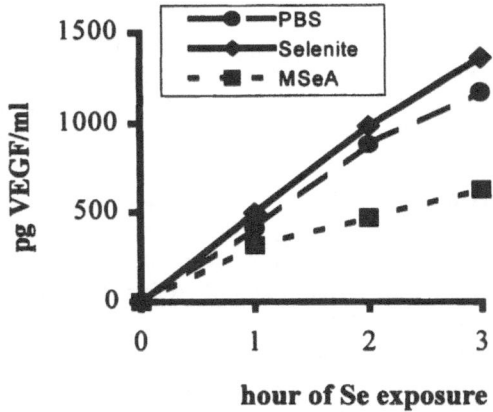

C. DNA fragmentation at 24 h

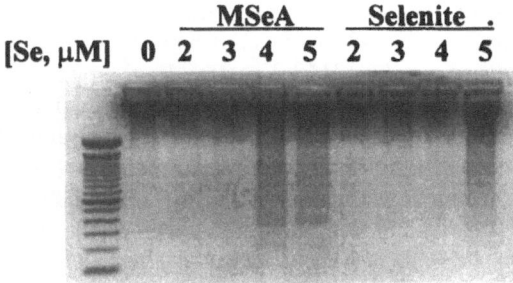

Figure 2. A: concentration-dependent effects of methylseleninic acid (MSeA, squares) and selenite (diamonds) on VEGF protein levels in DU 145 prostate carcinoma cell lysate after 6 h of exposure in serum-rich medium. Secreted VEGF in conditioned medium had similar response curves. B: acute time course of effects of exposure to Se as MSeA (5 μM, squares) or as sodium selenite (10 μM, diamonds) on VEGF expression in DU 145 cells. In these short-term exposure experiments, near-confluent DU 145 cells were treated with Se in 10% fetal bovine serum medium. Treatment in serum-free medium showed identical patterns of responses. VEGF was assayed with an enzyme-linked immunosorbent assay kit (R & D Systems, Minneapolis, MN). C: DNA nucleosomal fragmentation assay as an apoptosis marker after 24 h of exposure to MSeA or selenite (adherent and detached cells were combined for DNA extraction). (Modified from Ref. 6.)

served at a concentration (2 μM) that was twofold lower than that needed to induce significant apoptosis (≥ 4 μM). Chronic exposure of DU 145 cells to low-level MSeA mimicking a daily supplement regimen indicated a sustained inhibition of VEGF expression (6). Although selenite and MSeA were almost equipotent for inducing DNA apoptotic fragmentation (Fig. 2C), selenite did not inhibit VEGF expression by acute or chronic exposure. In support of the generality of this methyl Se-inhibitory effect on VEGF expression, two human breast cancer cell lines (MCF-7 and MDA-MB-468) showed the same Se metabolite specificity of inhibition irrespective of the serum level in the treatment medium (6). Taken together, these data support a primary and sustained methyl Se-specific inhibitory activity of serum-achievable Se on the expression of VEGF in cancer epithelial cells independently of cell death induction.

Methyl Se-Specific Inhibition of Endothelial MMP-2 Expression

As documented recently (6), treatment of HUVECs for 6 h with MSeA led to a concentration-dependent reduction of the secreted 72-kDa MMP-2 gelatinolytic activity in the conditioned medium (Fig. 3A). The inhibitory efficacy was remarkable, with an IC$_{50}$ of ~2 μM. Similarly, treatment with MSeCN resulted in a concentration-dependent decrease of MMP-2, and the inhibitory efficacy was comparable to that of MSeA (Fig. 3A). In contrast, treatment with hydrogen selenide precursors (up to 20 μM sodium selenite or 50 μM sodium selenide) did not significantly decrease MMP-2 in the conditioned medium (Fig. 3A).

Incubation of HUVEC-conditioned medium (containing secreted MMP-2) with all four Se forms in the test tube for 6 h at 37°C did not decrease the gelatinolytic activity (Fig. 3B), indicating that MSeA or MSeCN per se did not react directly with MMP-2 protein to inactivate its activity. The inhibitory action of these methylselenol precursor compounds was, therefore, dependent on cellular metabolism. Western blot analyses for the MMP-2 protein level in conditioned medium of HUVECs treated with MSeA indicated that a reduction in the MMP-2 protein level closely paralleled the observed loss of MMP-2 gelatinolytic activity (Fig. 3C), whereas selenite treatment had a minimal effect on the gelatinolytic activity and the MMP-2 protein level (Fig. 3C). The inhibitory effect of MSeA on MMP-2 was rather rapid, in that ~50% reduction was detected within 30 min of exposure (Fig. 3D). Together, these results indicate that methylselenol or its related monomethyl metabolites in the serum-achievable range (IC$_{50}$ ~2 μM) exert a rapid inhibitory effect on MMP-2 expression in vascular endothelial cells.

Methyl Se Potently Inhibited Angiogenic Factor-Driven Mitogenesis of Endothelial Cells Independently of Apoptosis Induction

We recently used a synchronized HUVEC model to more sensitively define the effect of methyl Se on angiogenic factor-stimulated mitogenesis (48). The bovine pituitary extract endothelial cell growth supplement (ECGS) was omitted from the complete medium for >48 h to arrest cells in the G$_0$/G$_1$ phase. Resumption of ECGS stimulation resulted in a >10-fold incorporation of [³H]thymidine within 24 h (Fig. 4A). MSeA treatment that was initiated 3 h before or simultaneously with ECGS stimulation decreased the ECGS-stimulated DNA synthesis dose dependently with an IC$_{50}$ of ~1 μM and completely blocked this parameter at 3 μM (Fig. 4A). The potent antimitogenic activity was independent of apoptosis induction, which required >5 μM MSeA after 30 h of exposure, as indicated by caspase-mediated cleavage of poly(ADP-ribose) polymerase and DNA fragmentation, both being biochemical hallmarks of apoptosis (Fig. 4B).

Summary and Discussion

The data reviewed above support a potential antiangiogenic activity of Se in vivo (5) and, furthermore, clearly contrast the differential effects of methylselenol vs. hydrogen selenide pools on multiple aspects of angiogenesis regulation (6). The in vitro data show that methyl Se potently inhibits ECGS-induced cell cycle progression of vascular endothelial cells to the S phase (IC$_{50}$ ~1 μM). In addition, methyl Se exerts an inhibitory activity on endothelial expression of MMP-2 (IC$_{50}$ ~2 μM) and on epithelial expression of VEGF (IC$_{50}$ ~2 μM). As reference values, the mean plasma Se concentration of subjects without Se supplementation in the trial of Clark et al. (1) was ~1.5 μM, and Se supplementation (200 μg/day as selenized yeast) brought the mean Se level to ~2.4 μM. Therefore, the methyl Se-specific antimitogenic activity and inhibitory activities on VEGF and MMP-2 proteins may be physiologically very pertinent for angiogenic switch regulation in early transformed lesions in vivo in the context of cancer chemoprevention, which aims at retarding and blocking the growth and progression of early lesions. Our data also show that methyl Se can induce apoptosis of vascular endothelial cells through caspase-mediated execution, but such a proapoptotic effect may be relevant only in a pharmacological context of Se exposure.

Integrating Endothelial and Epithelial Mechanisms for Cancer Prevention by Se

Before our work that was reviewed here, much mechanistic research had focused on how Se affected the cancer epithelial cells through antiproliferative and proapoptotic pathways (3,13). Because epithelial lesions do not exist in isolation in vivo but, instead, intimately interact with the stroma and vasculature, cancer prevention function is likely achieved through integrating the actions of Se on epithelial as well as nonepithelial targets. The physiochemistry of Se delivery to transformed epithelial cells in in vivo lesions may be a major determinant of the actual mechanism(s) as well as the processes that are invoked to regulate the growth and fate of the solid lesion. To this end, we speculated earlier (13) that Se delivery to epithelial cells in the avascular le-

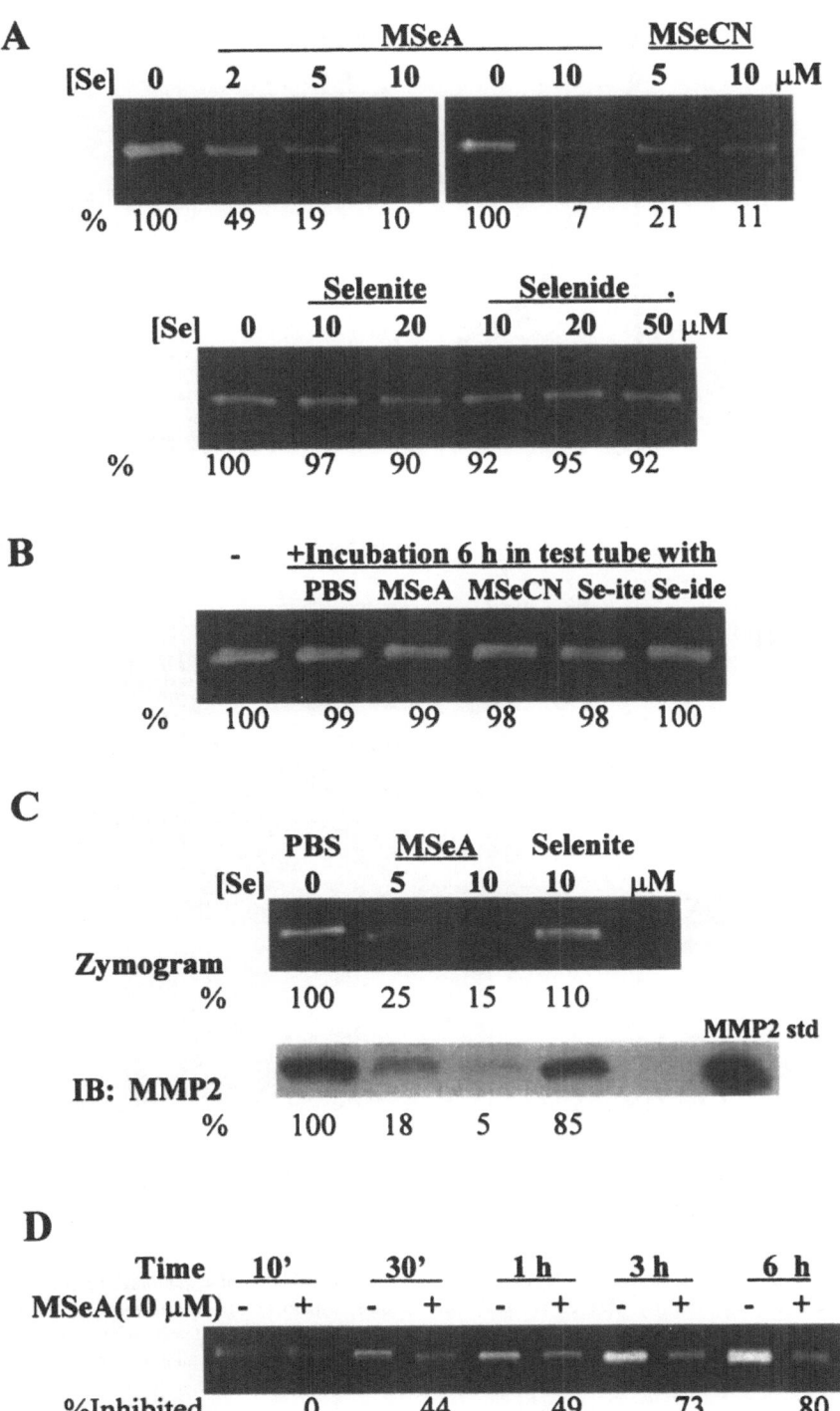

Figure 3. A: inhibitory specificity of selenium forms on human umbilical vein endothelial cell (HUVEC) MMP-2. Representative gelatin substrate gel zymographic analyses of MMP-2 were carried out in conditioned media of HUVECs treated for 6 h (in separate experiments) with MSeA, methylselenocyanate (MSeCN), sodium selenite, or sodium selenide in serum-free medium supplemented with endothelial cell growth supplement (ECGS, 100 μg/ml). Relative pixel density as percentage of control cells is shown below each lane. B: lack of MMP-2 inactivation by direct incubation of HUVEC-conditioned medium with Se compounds in test tubes at 37°C for 6 h. Each Se was added to 10 μM. C: immunoblot (IB) analyses of MMP-2 protein in conditioned media. HUVECs were treated with MSeA or selenite for 3 h. Aliquots of conditioned media were analyzed by zymography for gelatinolytic activity. Bulk of conditioned media was concentrated ~50-fold using Centricon spin filters and analyzed by Western blot. Pro-MMP-2 protein (10 ng) was used as standard. D: time course of MSeA (10 μM)-induced inhibition of MMP-2 expression in HUVECs. Aliquots of media taken at various time points were analyzed by zymography. Percent inhibition relative to phosphate-buffered saline (PBS)-treated control at each time point is shown below treated lanes. (Modified from Ref. 6.)

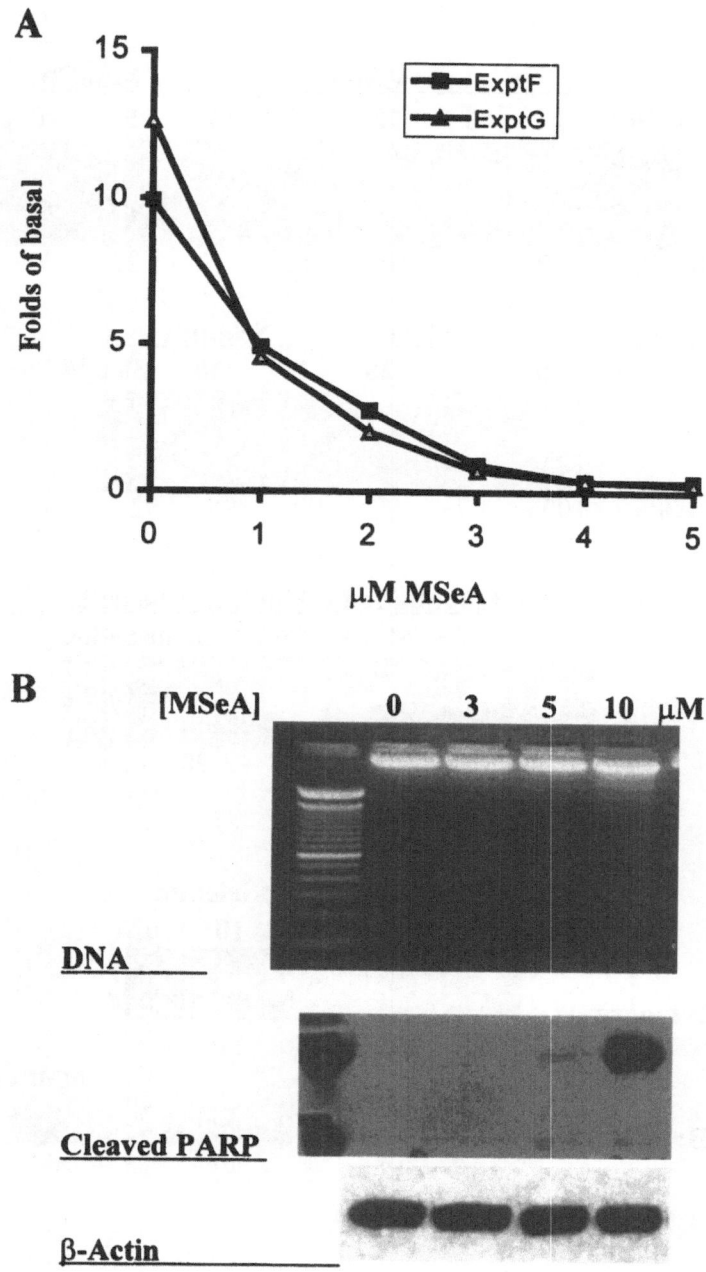

Figure 4. A: inhibitory effect of MSeA on ECGS-stimulated [^{3}H]thymidine incorporation into DNA of synchronized HUVECs. [^{3}H]thymidine was added at time of treatments with ECGS for 30 h. In *experiment F*, MSeA was added simultaneously with ECGS. In *experiment G*, MSeA was added 3 h before ECGS. B: biochemical markers of apoptosis. Detached and adherent cells were used for DNA extraction and for detection of cleaved poly(ADP-ribose) polymerase (PARP) in MSeA-exposed HUVECs by Western blot. β-Actin was reprobed to indicate evenness of loading of protein extract from each treatment.

sions may follow a concentration gradient similar to oxygen tension, which has been known to decline precipitously as the distance to the nearby microvessel increases, resulting in a hypoxic state in the interior of such lesions (36) (Fig. 5A). Should such a declining gradient exist for Se, a "conditional Se deficiency" state may be created inside, expanding microscopic avascular epithelial lesions even when the Se supply is sufficient to saturate selenoprotein activities in the serum or normal tissues.

This model predicts that more Se is required to enrich Se metabolite pools within the avascular lesions in order to elicit antimitogenic and proapoptotic pathways in the transformed epithelial cells. Furthermore, more Se is required to support the activity of key selenoproteins such as thioredoxin reductases and Se-glutathione peroxidases in cells within the epithelial lesions so as to reregulate their transformation status/physiology. This model may warrant a reevaluation of the current paradigm that discounts the likelihood of involvement of Se-glutathione peroxidases and other selenoproteins for the chemopreventive activity of Se (3,4). Consistent with this speculation, the prostate cancer-preventive effect of 200 μg of Se as Se-yeast in the trial of Clark

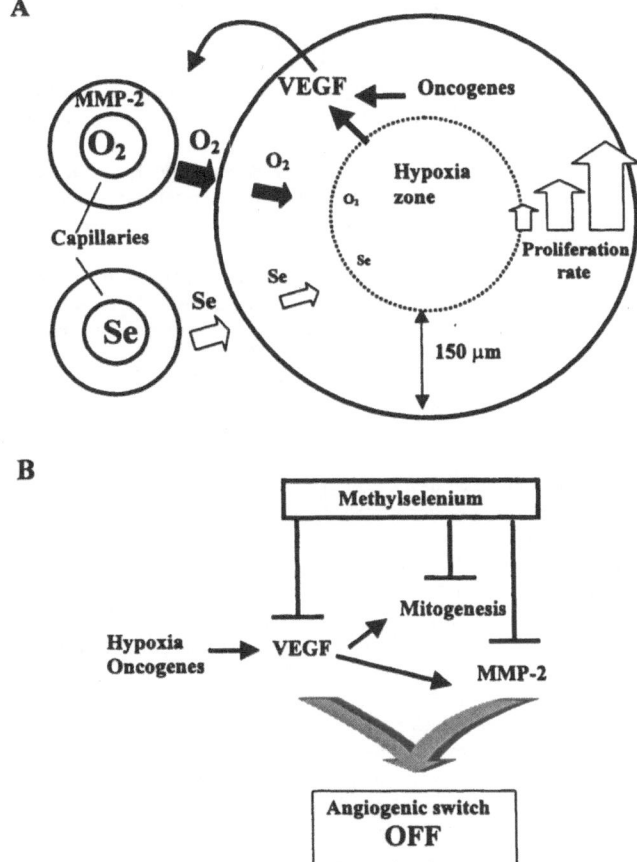

A

B

Figure 5. A: a hypothetical model of Se delivery to epithelial cells within an avascular lesion analogous to oxygen diffusion. A conditional Se deficiency is speculated in cells that are also hypoxic. Increased VEGF expression due to hypoxia and oncogenic mutations may be a primary stimulatory signal for angiogenesis switch in early lesions. B: methylselenol-specific inhibitory effects on angiogenic switch. In vitro data support an inhibition of cancer epithelial expression of VEGF, endothelial mitogenesis, and MMP-2 expression, turning off the switch.

et al. appeared to be most prominent in patients with entry plasma Se in the lowest tertile (49).

This model also highlights the need for investigations that incorporate hypoxia as a feature of the solid lesions in evaluation of the efficacy of Se to induce growth arrest and cell death responses as well as angiogenesis regulation. Hypoxia is known to affect cellular energy metabolism and cellular redox states (36). It would be important to address whether a hypoxic state alters the proapoptotic efficacy of the different pools of Se metabolites. Hypoxia is a potent stimulus for the angiogenic switch by stimulating VEGF expression (37). An inhibitory activity by methyl Se on the ability of the epithelial lesions to produce this angiogenic factor represents a metabolite-specific avenue through which Se can regulate angiogenesis and carcinogenesis (Fig. 5B).

In addition to the Se delivery considerations in epithelial lesions, strong arguments can be made in support of antiendothelial and antiangiogenesis actions as primary effectors of cancer prevention by Se metabolites. The angiogenic microenvironment is likely conducive for selective targeting of stimulated vascular endothelial cells by blood-borne Se metabolites. Several physical and biological features of endothelial cells and the newly formed vessels stand out in contrast to tumor cells: 1) as lining of blood vessels, endothelial cells are the first line of exposure to blood Se; 2) vascular endothelial cells in nonangiogenic environment are essentially quiescent and protected by pericytes and smooth muscle cells; and 3) angiogenic factor-stimulated endothelial cells undergo mitogenic and other responses to form new sprouts, which are leaky and not as protected as are existing vessels (50). Sustained angiogenic stimulation provided by the transformed cells in the expanding lesions is essential for continued angiogenesis within a tumor, in contrast to normal angiogenesis during menstrual cycle or wound healing, for example, where expression of angiogenic regulators (positive and negative factors) is tightly controlled (51). This fact is clearly articulated by the saying that "cancer is a wound that never heals." Such a difference may provide a plausible biological basis for selectivity against malignancy through suppressing tumor angiogenesis. Our data suggest that, in a chemoprevention context, the increased circulating and tissue levels of Se metabolites, especially the monomethyl Se pool, can exert a direct antimitogenic activity on the vascular endothelial cells (Fig. 5B) and perhaps even induce apoptosis. Because endothelial cells are much less susceptible to mutagenesis than are epithelial cells, the wild-type p53 tumor suppressor status of the endothelial cells can ensure p53-dependent growth arrest and apoptosis. The methylselenol specific inhibitory activity on endothelial MMP-2 expression (Fig. 5B) will further dampen the endothelial responsiveness to angiogenic stimulation. These direct antiendothelial effects combined with inhibitory action on epithelial expression of VEGF can serve as a powerful control for the angiogenic switch and, hence, carcinogenesis (Fig. 5B).

The new paradigm advocated here emphasizes the nonepithelial and epithelial targets for Se action. The angiogenic switch regulatory activities described for the methyl Se pool may provide a mechanism for novel biochemical and molecular markers of cancer prevention by Se. The specific nature of the angiogenic switch control mechanisms (52) and a difference in the ability to generate methylselenol may underscore organ site specificity of cancer-preventive activity of Se, as has been documented in the trial of Clark et al. (prostate, colon, and lung vs. skin). These issues merit further investigation.

Acknowledgments and Notes

Data cited in this review represent team efforts of many individuals, including Zaisen Wang, Weiqin Jiang (AMC Cancer Research Center), and our collaborators Howard Ganther and Clement Ip. Recent work in the Lu laboratory has been supported in part by grants from the American Institute for Cancer Research, the Department of Defense, and the National Cancer Institute. Address correspondence to Dr. Junxuan Lu, AMC Cancer Research Center, Center for Cancer Causation and Prevention, 1600

Pierce St., Denver, CO 80214. Phone: 303-239-3348. FAX: 303-239-3560. E-mail: luj@amc.org.

Submitted 30 April 2001; accepted in final form 4 May 2001.

References

1. Clark LC, Combs GF Jr, Turnbull BW, Slate EH, Chalker DK, et al.: Effects of selenium supplementation for cancer prevention in patients with carcinoma of the skin: a randomized controlled trial. Nutritional Prevention of Cancer Study Group. *JAMA* **276**, 1957–1963, 1996.

2. Yu SY, Zhu YJ, and Li WG: Protective role of selenium against hepatitis B virus and primary liver cancer in Qidong. *Biol Trace Elem Res* **56**, 117–124, 1997.

3. Ip C: Lessons from basic research in selenium and cancer prevention. *J Nutr* **128**, 1845–1854, 1998.

4. Combs GF Jr and Gray WP: Chemopreventive agents: selenium. *Pharmacol Ther* **79**, 179–192, 1998.

5. Jiang C, Jiang W, Ip C, Ganther H, and Lu JX: Selenium-induced inhibition of angiogenesis in mammary cancer at chemopreventive levels of intake. *Mol Carcinog* **26**, 213–225, 1999.

6. Jiang C, Ganther H, and Lu JX: Monomethyl selenium-specific inhibition of MMP-2 and VEGF expression: implications for angiogenic switch regulation. *Mol Carcinog* **29**, 236–250, 2000.

7. Ip C and Ganther HE: Activity of methylated forms of selenium in cancer prevention. *Cancer Res* **50**, 1206–1211, 1990.

8. Ip C, Hayes C, Budnick RM, and Ganther HE: Chemical form of selenium, critical metabolites, and cancer prevention. *Cancer Res* **51**, 595–600, 1991.

9. Lu JX, Jiang C, Kaeck M, Ganther H, Vadhanavikit S, et al.: Dissociation of the genotoxic and growth inhibitory effects of selenium. *Biochem Pharmacol* **50**, 213–219, 1995.

10. Lu JX, Pei H, Ip C, Lisk D, Ganther H, et al.: Effect of an aqueous extract of selenium-enriched garlic on in vitro markers and in vivo efficacy in cancer prevention. *Carcinogenesis* **17**, 1903–1907, 1996.

11. Kaeck M, Lu JX, Strange R, Ip C, Ganther HE, et al.: Differential induction of growth arrest inducible genes by selenium compounds. *Biochem Pharmacol* **53**, 921–926, 1997.

12. Sinha R and Medina D: Inhibition of cdk2 kinase activity by methylselenocysteine in synchronized mouse mammary epithelial tumor cells. *Carcinogenesis* **18**, 1541–1547, 1997.

13. Lu JX: Apoptosis and angiogenesis in cancer prevention by selenium. In *Nutrition and Cancer Prevention: New Insights Into the Role of Phytochemicals*. New York: Kluwer Academic/Plenum, 2000, pp 131–145.

14. Jiang C, Wang Z, Ganther H, and Lu JX: Caspases as key executors of methyl selenium-induced apoptosis (anoikis) of DU-145 human prostate cancer cells. *Cancer Res* **61**, 3062–3070, 2001.

15. Folkman J: The role of angiogenesis in tumor growth. *Semin Cancer Biol* **3**, 65–71, 1992.

16. Hanahan D and Folkman J: Patterns and emerging mechanisms of the angiogenic switch during tumorigenesis. *Cell* **86**, 353–364, 1996.

17. Bouck N, Stellmach V, and Hsu SC: How tumors become angiogenic. *Adv Cancer Res* **69**, 135–174, 1996.

18. Zetter BR: Angiogenesis and tumor metastasis. *Annu Rev Med* **49**, 407–424, 1998.

19. Leung DW, Cachianes G, Kuang WJ, Goeddel DV, and Ferrara N: Vascular endothelial growth factor is a secreted angiogenic mitogen. *Science* **246**, 1306–1309, 1989.

20. Keck PJ, Hauser SD, Krivi G, Sanzo K, Warren T, et al.: Vascular permeability factor, an endothelial cell mitogen related to PDGF. *Science* **246**, 1309–1312, 1989.

21. Grunstein J, Roberts WG, Mathieu-Costello O, Hanahan D, and Johnson RS: Tumor-derived expression of vascular endothelial growth factor is a critical factor in tumor expansion and vascular function. *Cancer Res* **59**, 1592–1598, 1999.

22. Brown LF, Guidi AJ, Tognazzi K, and Dvorak HF: Vascular permeability factor/vascular endothelial growth factor and vascular stroma formation in neoplasia. Insights from in situ hybridization studies. *J Histochem Cytochem* **46**, 569–575, 1998.

23. Guidi AJ, Abu-Jawdeh G, Tognazzi K, Dvorak HF, and Brown LF: Expression of vascular permeability factor (vascular endothelial growth factor) and its receptors in endometrial carcinoma. *Cancer* **78**, 454–460, 1996.

24. Guidi AJ, Schnitt SJ, Fischer L, Tognazzi K, Harris JR, et al.: Vascular permeability factor (vascular endothelial growth factor) expression and angiogenesis in patients with ductal carcinoma in situ of the breast. *Cancer* **80**, 1945–1953, 1997.

25. Abu-Jawdeh GM, Faix JD, Niloff J, Tognazzi K, Manseau E, et al.: Strong expression of vascular permeability factor (vascular endothelial growth factor) and its receptors in ovarian borderline and malignant neoplasms. *Lab Invest* **74**, 1105–1115, 1996.

26. Fukumura D, Xavier R, Sugiura T, Chen Y, Park EC, et al.: Tumor induction of VEGF promoter activity in stromal cells. *Cell* **94**, 715–725, 1998.

27. Carmeliet P, Ferreira V, Breier G, Pollefeyt S, Kieckens L, et al.: Abnormal blood vessel development and lethality in embryos lacking a single VEGF allele. *Nature* **380**, 435–439, 1996.

28. Ferrara N, Carver-Moore K, Chen H, Dowd M, Lu L, et al.: Heterozygous embryonic lethality induced by targeted inactivation of the VEGF gene. *Nature* **380**, 439–442, 1996.

29. Zhang HT, Craft P, Scott PA, Ziche M, Weich HA, et al.: Enhancement of tumor growth and vascular density by transfection of vascular endothelial cell growth factor into MCF-7 human breast carcinoma cells. *JNCI* **87**, 213–219, 1995.

30. Aonuma M, Saeki Y, Akimoto T, Nakayama Y, Hattori C, et al.: Vascular endothelial growth factor overproduced by tumour cells acts predominantly as a potent angiogenic factor contributing to malignant progression. *Int J Exp Pathol* **80**, 271–281, 1999.

31. Borgstrom P, Hillan KJ, Sriramarao P, and Ferrara N: Complete inhibition of angiogenesis and growth of microtumors by anti-vascular endothelial growth factor neutralizing antibody: novel concepts of angiostatic therapy from intravital videomicroscopy. *Cancer Res* **56**, 4032–4039, 1996.

32. Benjamin LE and Keshet E: Conditional switching of vascular endothelial growth factor (VEGF) expression in tumors: induction of endothelial cell shedding and regression of hemangioblastoma-like vessels by VEGF withdrawal. *Proc Natl Acad Sci USA* **94**, 8761–8766, 1997.

33. Borgstrom P, Bourdon MA, Hillan KJ, Sriramarao P, and Ferrara N: Neutralizing anti-vascular endothelial growth factor antibody completely inhibits angiogenesis and growth of human prostate carcinoma micro tumors in vivo. *Prostate* **35**, 1–10, 1998.

34. Meeson AP, Argilla M, Ko K, Witte L, and Lang RA: VEGF deprivation-induced apoptosis is a component of programmed capillary regression. *Development* **126**, 1407–1415, 1999.

35. Benjamin LE, Golijanin D, Itin A, Pode D, and Keshet E: Selective ablation of immature blood vessels in established human tumors follows vascular endothelial growth factor withdrawal. *J Clin Invest* **103**, 159–165, 1999.

36. Brown JM and Giaccia AJ: The unique physiology of solid tumors: opportunities (and problems) for cancer therapy. *Cancer Res* **58**, 1408–1416, 1998.

37. Levy AP, Levy NS, Wegner S, and Goldberg MA: Transcriptional regulation of the rat vascular endothelial growth factor gene by hypoxia. *J Biol Chem* **270**, 13333–13340, 1995.

38. Rak J, Mitsuhashi Y, BayKo L, Filmus J, Shirasawa S, et al.: Mutant *ras* oncogenes upregulate VEGF/VPF expression: implications for induction and inhibition of tumor angiogenesis. *Cancer Res* **55**, 4575–4580, 1995.

39. Grugel S, Finkenzeller G, Weindel K, Barleon B, and Marme D: Both v-Ha-Ras and v-Raf stimulate expression of the vascular endothelial growth factor in NIH 3T3 cells. *J Biol Chem* **270**, 25915–25919, 1995.

40. Mazure NM, Chen EY, Yeh P, Laderoute KR, and Giaccia AJ: Oncogenic transformation and hypoxia synergistically act to modulate vascular endothelial growth factor expression. *Cancer Res* **56**, 3436–3440, 1996.

41. Arbiser JL, Moses MA, Fernandez CA, Ghiso N, Cao Y, et al.: Oncogenic H-*ras* stimulates tumor angiogenesis by two distinct pathways. *Proc Natl Acad Sci USA* **94**, 861–866, 1997.

42. Akagi Y, Liu W, Zebrowski B, Xie K, and Ellis LM: Regulation of vascular endothelial growth factor expression in human colon cancer by insulin-like growth factor-I. *Cancer Res* **58**, 4008–4014, 1998.

43. Ryuto M, Ono M, Izumi H, Yoshida S, Weich HA, et al.: Induction of vascular endothelial growth factor by tumor necrosis factor-α in human glioma cells: possible roles of SP-1. *J Biol Chem* **271**, 28220–28228, 1996.

44. Deroanne CF, Hajitou A, Calberg-Bacq CM, Nusgens BV, and Lapiere CM: Angiogenesis by fibroblast growth factor 4 is mediated through an autocrine up-regulation of vascular endothelial growth factor expression. *Cancer Res* **57**, 5590–5597, 1997.

45. Coussens LM and Werb Z: Matrix metalloproteinases and the development of cancer. *Chem Biol* **3**, 895–904, 1996.

46. Itoh T, Tanioka M, Yoshida H, Yoshioka T, Nishimoto H, et al.: Reduced angiogenesis and tumor progression in gelatinase A-deficient mice. *Cancer Res* **58**, 1048–1051, 1998.

47. Hiraoka N, Allen E, Apel IJ, Gyetko MR, and Weiss SJ: Matrix metalloproteinases regulate neovascularization by acting as pericellular fibrinolysins. *Cell* **95**, 365–377, 1998.

48. Wang Z, Jiang C, Ganther H, and Lu JX: Anti-mitogenic and pro-apoptotic activities of methylseleninic acid in vascular endothelial cells and associated effects on PI3K-AKT, ERK, JNK and p38MAPK signaling. *Cancer Res* **61**, 7171–7178, 2001.

49. Dalkin BW, Lillico AJ, Reid ME, Jacobs ET, Combs GF, et al.: Selenium and chemoprevention against prostate cancer: an update on the Clark results (abstr). *Proc Am Assoc Cancer Res 2001*, p 460.

50. Dvorak HF: VPF/VEGF and the angiogenic response. *Semin Perinatol* **24**, 75–78, 2000.

51. Torry RJ and Rongish BJ: Angiogenesis in the uterus: potential regulation and relation to tumor angiogenesis. *Am J Reprod Immunol* **27**, 171–179, 1992.

52. Eberhard A, Kahlert S, Goede V, Hemmerlein B, Plate KH, et al.: Heterogeneity of angiogenesis and blood vessel maturation in human tumors: implications for antiangiogenic tumor therapies. *Cancer Res* **60**, 1388–1393, 2000.

NUTRITION AND CANCER, 40(1), 74–77
Copyright © 2001, Lawrence Erlbaum Associates, Inc.

Larry Clark's Legacy: Randomized Controlled, Selenium-Based Prostate Cancer Chemoprevention Trials

James R. Marshall

Abstract: Several important clinical trials under way at the Arizona Cancer Center seek to build on the results of Clark's 1996 study of selenium and decreased risk of prostate cancer. Those results, an unanticipated end point of a clinical trial, suggest that selenium has significant preventive power. The studies under way involve continued follow-up of the study cohort that generated the 1996 results and trials of selenium among men with negative biopsies, men with high-grade prostatic intraepithelial neoplasia, men with prostate cancer treated with selenium before prostatectomy, and men with prostate cancer who have chosen watchful waiting rather than active intervention. These studies promise important opportunities to validate Clark's original results.

Introduction

Several prostate cancer chemoprevention studies under way at the University of Arizona were inspired largely by the results of the trial of Clark et al. (1) of selenium among men and women who had been treated for basal or squamous cell skin cancer. The study of Clark et al. was originally designed to prevent the recurrence of skin cancer among these individuals. Selenium was found not to decrease the risk of skin cancer recurrence. However, individuals in the experimental group, treated with selenium at 200 µg/day in baker's yeast, experienced sizable decreases in the risks of primary cancers of the colon and rectum, prostate, and lung. Prostate cancer incidence decreased by a striking 65%. These results are important; the study was an experiment, so experimental subjects were exposed to much higher levels of selenium over the trial's duration than were controls. Subjects were, on average, treated for 4.5 yr. The experimental design provides excellent documentation that selenium supplementation had a substantial effect on blood levels; within 1 yr, experimental subject blood levels had reached a new equilibrium, with those levels approximately doubled. In addition, because assignment to the treatment group was randomized, the probability of confounding of the treatment

effect by other factors that might have altered risk of prostate cancer was decreased.

Additional studies of the prostate cancer effect are necessary to confirm the trial results and to extend them to other populations. The greatly decreased prostate cancer observed in the study was not originally anticipated, and this decreases our confidence in the effect (2). We have designed and initiated four new double-blind, randomized trials to confirm the observation of Clark et al. (1,2) that selenium treatment decreased the risk of prostate cancer. These trials will be conducted among men at various degrees of elevated prostate cancer risk.

The results of this important trial demonstrated how little we understand regarding the mechanisms by which selenium might affect the risk of cancer. The trial left a number of scientific issues to be considered. One of these is the content of the high-selenium baker's yeast used in the trial of Clark et al. (1,2). They used a selenized yeast that was commercially available. Part of the attractiveness of this agent was that it was readily available as a food supplement. The disadvantage was that it had not been analyzed for the different forms of selenium with as much detail as is desirable. This selenized yeast contains several selenium compounds, of which only a few have been identified: L-selenomethionine, selenocysteine, Se-methylselenocysteine, selenoethionine, selenoglutathione, selenodiglutathione, and selenite are believed to be the most prominent and active (3,4). Recent evaluations of this selenized yeast reveal that most of the selenium is in the form of L-selenomethionine (5). The mechanisms by which selenomethionine might protect are not completely understood. Selenomethionine is metabolized and stored in the same manner as methione (4). Catabolism of the selenomethionine then yields selenocysteine, which yields several selenoproteins and methylselenol. It has been experimentally documented (4) that methylselenol may be highly active in blocking a number of carcinogenic pathways. We are presently involved in a series of studies designed to provide more detailed speciation of the high-selenium yeast that had been used by Clark et al. In addition, the mechanisms by which any chemopreventive effects might be realized are not known. Ip recently pointed out that,

---

The author is affiliated with the Arizona Cancer Center, University of Arizona, Tucson, AZ 84724.

in the unfolding of research with selenium, epidemiological and clinical trial results are ahead of basic science understanding (4). A great deal of basic science evidence has been accumulated, and this science suggests a number of mechanisms by which selenium may have a protective effect against cancer (4). We are not able to convincingly confirm or dismiss any of these possible causal mechanisms.

The four studies under way at the Arizona Cancer Center seek to extend our understanding. These randomized, placebo-controlled, double-blind trials offer opportunities to confirm the results of Clark et al. (1,2) and to explore the mechanisms by which selenium compounds might work. These trials, involving intervention at different phases of the development of clinical prostate cancer, complement one another. The use of a unitary, as opposed to a complex, mixture is explored; the treatment in one trial is pure selenomethionine; the other trials use high-selenium baker's yeast. As mentioned, it is hypothesized that selenomethionine, one of the major compounds in selenized yeast, is responsible for the protective effects observed by Clark et al. This trial will provide an important opportunity to evaluate that hypothesis, inasmuch as none of the other forms of selenium in selenized yeast will be present. This offers a limited opportunity to explore whether the primary compound in selenized yeast, selenomethionine, has effects that are different from those of the full range of selenium compounds present in selenized yeast.

The first study is being conducted among men who have been biopsied for prostate cancer but whose biopsy was negative: the negative biopsy trial. The second trial involves treatment with selenium of men with high-grade prostatic intraepithelial neoplasia (HGPIN). The third trial involves selenium treatment of men with localized prostate cancer before prostatectomy. The fourth trial, watchful waiting, will test selenium as a chemotherapeutic agent among men with confirmed prostate cancer. These selenium intervention studies range from evaluations of selenium among high-risk individuals who have not been diagnosed with prostate cancer to attempts to alter the course of clearly invasive prostate cancer. All will evaluate changes in prostate-specific antigen (PSA), and all are restricted to individuals who are not taking PSA-altering drugs.

Negative Biopsy Trial

Men with persistently elevated PSA are at substantially elevated risk of prostate cancer (6) and, thus, represent a high-priority target for prostate cancer chemoprevention studies. A recent negative biopsy of the prostate will help ensure, to the degree possible, that these men do not already have clinical prostate cancer. As many as one-fourth of these men will have early prostate cancer, missed by the biopsy, so they represent a group that is less homogeneous than desired. In general, however, men whose biopsies are negative are likely to have minimal disease, and it is possible that selenium may be active against the progression of cancer at

this minimal disease stage. The randomization procedure will help ensure that they are equally distributed in the treatment groups. Thus 700 men with negative biopsies but elevated PSA will be randomized to placebo or 200 or 400 μg of selenium in high-selenium baker's yeast. Follow-up is expected to be up to 57 mo. Outcomes anticipated include diagnosed prostate cancer and the rate of rise in the PSA before the end of treatment and follow-up. This trial will evaluate the ability of selenium, first, to decrease the incidence of clinical prostate cancer and, second, to halt or slow the preclinical progression of prostate cancer. This trial, funded by the National Cancer Institute (NCI), should be directly generalizable to those patients at high risk of prostate cancer but who have not yet been diagnosed. So far, 210 men have been randomized to this trial.

HGPIN Trial

Men with HGPIN are at substantially elevated risk of subsequent prostate cancer (7,8). However, such men are not treated by surgery or by irradiation. The identification of a compound that would decrease the risk of cancer development among these men could be of great value. This trial, funded by the NCI, is being conducted through the Southwest Oncology Group. Altogether, 470 men with HGPIN, who have had two or more biopsies that indicate no prostate cancer, will be assigned to placebo or to 200 μg of selenium daily as selenomethionine. They will be followed for ≥3 yr. Those who have not had additional biopsies will be scheduled for biopsy at the end of the follow-up period. The primary end point in this study is the diagnosis of biopsy-proven prostate cancer. Secondary end points, biomarkers of change, include apoptosis and proliferation. Machine-vision imaging will also be used to evaluate change in nuclear characteristics and degradation of basal cell integrity in the glands and ducts (9,10). This study will evaluate the activity of selenium immediately before neoplastic growth becomes transformed into invasive growth. An important possible limitation of the study is that a small percentage of these men will have prostate cancer missed by the biopsies. The use of multiple biopsies at baseline, however, should minimize this problem. This study has registered 85 subjects.

Preprostatectomy Trial

This trial is funded by the Department of Defense to enroll 110 men with biopsy-diagnosed, localized prostate cancer who elect treatment with prostatectomy and agree to treatment with selenium or placebo during the period before surgery. Subjects will be assigned to placebo or 200 or 400 μg of selenium per day as high-selenium baker's yeast. Subjects will be evaluated primarily during the 6- to 8-wk period between biopsy and prostatectomy. This short study period is necessitated by the fact that men who have been diagnosed with prostate cancer are usually anxious to be treated as soon as possible. The most common clinical practice is to allow 6–8 wk for biopsy-induced inflammation to subside and then to proceed with prostatectomy. No run-in period for patient compliance is included in the trial because of the relatively short time available for treatment. Study end points include immunohistochemical biomarkers of change in healthy tissue, including proliferation, apoptosis, thioredoxin, thioredoxin reductase, and glutathione peroxidase. These selenoproteins are regarded as highly unlikely to be the means by which selenium supplementation could alter prostate cancer risk (11). The levels of these could well reflect the extent of antioxidant tissue defense within the prostate. Fresh prostate tissue will be obtained from prostatectomy for evaluation of selenium levels. This clinical trial will assess whether selenium administration can change the biochemistry of the prostate within a very brief period. It will allow an evaluation of whether what are believed to be important mechanisms of carcinogenesis, proliferation and apoptosis and the balance between them, can be altered by short-term treatment with selenium. At this time, 55 people have been recruited to this study.

Watchful Waiting Trial

There is evidence that, for men with localized disease and low-grade tumors and a life expectancy <10 yr, watchful waiting is a reasonable alternative to aggressive treatment (12,13). A nontoxic chemopreventive agent such as selenium would be an attractive adjuvant to watchful waiting. Thus, with NCI funding, 264 men with localized prostate cancer (PSA <50), who have chosen not to undergo prostatectomy, irradiation, or hormonal treatment, will be treated with placebo or 200 or 800 µg of selenium per day as yeast. In a separate but related study funded by other sources, 30 men were randomized to receive 1,600 or 3,200 µg of selenium per day in yeast. Although no overt toxicity was observed in this arm of the trial, it was halted because of fears of possible long-term adverse effects. In the event of massive overdosing with selenium, kidney and liver damage can be seen; with microgram dosage, however, the only observed toxicity is garlic breath and nail and hair brittleness. We are presently evaluating the results of this small trial. Patients will be followed for up to 4 yr. The end points of this study include selenium toxicity, rates of increase in PSA, need to commence hormonal treatment, and development of regional and distant metastases. This study is among patients who have declined therapy and could thus be considered an evaluation of the therapeutic use of selenium. The principal advantage of this study is its minimizing of the bias that could be introduced by including patients without prostate cancer. One of the most important outcomes of this study will be the evidence accrued on the impact of pharmacological doses of selenium. The dose of 800 µg of selenium per day is approximately four times the dose used by Clark et al. (1) for the major arm of their original study. The 400 and 800 µg/day doses are supraphysiological and will contribute considerable information regarding evidence of selenosis with long-term use. To date, no evidence of toxicity has been observed among the total of 85 patients recruited.

Summary

These double-blind, randomized trials under way at the Arizona Cancer Center build on the pioneering results reported by Clark et al. (1). The studies will evaluate the activity of selenium at several points along a continuum ranging from short-term effects on healthy and cancerous prostatic tissue in men with diagnosed cancer, to long-term effects on healthy and premalignant tissue in men with HGPIN, to long-term effects on healthy tissue in high-risk men with negative biopsy, to long-term effects on cancerous tissue in men with frank cancer. These complementary studies offer important opportunities to replicate and extend the results of Clark et al. They also offer an opportunity for preliminary evaluation of mechanisms by which selenium treatment could result in the slower development or progression of prostate cancer. Larry Clark realized, well before most students of prevention, the importance of the randomized-controlled trial as a means of identifying preventive strategies. We at the Arizona Cancer Center intend to build on the important foundations established by Dr. Clark.

An important impediment to recruitment to these trials is that a substantial proportion of potential patients are already taking large doses of selenium. That these trials are legitimate, however, requires that the efficacy of selenium cannot be regarded as established. Colditz (2) emphasized that even the promising results of the initial trial of Clark et al. do not legitimate large-scale use of selenium as a chemopreventive agent. A large prostate cancer prevention trial, to be conducted among men at average risk, is about to begin. This trial, the Selenium and E Chemoprevention Trial, will evaluate vitamin E and selenium, considered in a 2 × 2 factorial study design, among 32,800 men. The trial will be led by the Southwest Oncology Group and the Veterans Administration. It will also recruit patients among the Cancer and Leukemia Group B and the Eastern Cooperative Oncology Group. We have proposed a large general population trial that will have substantial statistical power for effects more modest than those observed in the original trial of Clark et al. Although the focus of this trial will be prostate cancer prevention, it will also be open to women; it will be important to evaluate definitively the impact of selenium supplementation among women as well as men. We are in the process of seeking additional grant support to expand this trial.

Acknowledgments and Notes

This work was supported by National Cancer Institute Grants 1 UO1 CA-77178, 5 UO1 CA-79080-03, and 1RO1 CA-77789-02. Address correspondence to James R. Marshall, Ph.D., Cancer Prevention and Control,

Arizona Cancer Center, University of Arizona, PO Box 245024, Tucson, Arizona 84724-5024.

Submitted 13 November 2000; accepted in final form 11 January 2001.

References

1. Clark LC, Combs GF Jr, Turnbull BW, Slate EH, Chalker DK, et al.: Effects of selenium supplementation for cancer prevention in patients with carcinoma of the skin: a randomized controlled trial. Nutritional Prevention of Cancer Study Group. *JAMA* **276,** 1957–1963, 1996. [Erratum. *JAMA* **277,** 21 May 1997, p 1520.]

2. Colditz GA: Selenium and cancer prevention. Promising results indicate further trials required. *JAMA* **276,** 1984–1985, 1996.

3. Neve J: Human selenium supplementation as assessed by changes in blood selenium concentration and glutathione peroxidase activity. *J Trace Elem Med Biol* **9,** 65–73, 1995.

4. Ip C: Lessons from basic research in selenium and cancer prevention. *J Nutr* **128,** 1845–1854, 1998.

5. Patterson BH and Levander OA: Naturally occurring selenium compounds in cancer chemoprevention trials: a workshop summary. *Cancer Epidemiol Biomarkers Prev* **6,** 63–69, 1997.

6. Gann PH, Hennekens CH, and Stampfer MJ: A prospective evaluation of plasma prostate-specific antigen for detection of prostatic cancer. *JAMA* **273,** 289–294, 1995.

7. Bostwick DG and Brawer MK: Prostatic intra-epithelial neoplasia and early invasion in prostate cancer. *Cancer* **59,** 788–794, 1987.

8. Sakr WA and Grignon DJ: Prostatic intraepithelial neoplasia and atypical adenomatous hyperplasia: relationship to pathologic parameters, volume and spatial distribution of carcinoma of the prostate. *Anal Quant Cytol Histol* **20,** 417–423, 1998.

9. Bartels PH, da Silva VD, Montironi R, Hamilton PW, Thompson D, et al.: Chromatin texture signatures in nuclei from prostate lesions. *Anal Quant Cytol Histol* **20,** 407–416, 1998.

10. Bartels PH, Montironi R, Thompson D, Vaught L, and Hamilton PW: Statistical histometry of the basal cell/secretory cell bilayer in prostatic intraepithelial neoplasia. *Anal Quant Cytol Histol* **20,** 381–388, 1998.

11. Berggren M, Gallegos A, Gasdaska J, and Powis G: Cellular thioredoxin reductase activity is regulated by selenium. *Anticancer Res* **17,** 3377–3380, 1997.

12. Chodak GW, Thisted RA, Gerber GS, Johansson JE, Adolfsson J, et al.: Results of conservative management of clinically localized prostate cancer. *N Engl J Med* **330,** 242–248, 1994.

13. Fleming C, Wasson JH, Albertsen PC, Barry MJ, and Wennberg JE: A decision analysis of alternative treatment strategies for clinically localized prostate cancer. Prostate Patient Outcomes Research Team. *JAMA* **269,** 2650–2658, 1993.

NUTRITION AND CANCER, *40*(1), 78

ANNOUNCEMENTS

VEGETARIAN NUTRITION

The Fourth International Congress on Vegetarian Nutrition will be held in Loma Linda, CA on 17–22 March 2002. *Information:* Department of Nutrition, School of Public Health, Loma Linda University, Loma Linda, CA 92350. E-mail: ICVN@sph.llu.edu.

NUTRITION GOALS
FOR ASIA-VISION 2020

The IX Asian Congress of Nutrition will be held on 23–27 February 2003 in New Delhi, India. The Congress is sponsored by the Federation of Asian Nutrition Societies, Nutrition Society of India, and Nutrition Foundation of India. *Information:* Dr. C. Gopalan, Nutrition Foundation of India, C-13 Qutab Institutional Area, New Delhi 110016, India. Phone: 91-11-6857814 or 6962615. FAX: 91-11-6560106 or 6857814. E-mail: acn2003@yahoo.com. Website: www.acn2003india.net.

INSTRUCTIONS FOR AUTHORS

Lawrence Erlbaum Associates, Inc. welcomes manuscripts of the following types for publication in *NUTRITION AND CANCER: AN INTERNATIONAL JOURNAL*.

1) Original papers containing results of experimental, clinical, or statistical studies that are timely and well documented.
2) Letters to the Editor that deal with issues of importance to researchers in the field of nutrition and cancer. Experimental data should be the minimal amount required for adequate understanding.
3) Reviews on subjects of importance to researchers in the field of nutrition and cancer.
4) Brief reports of meetings and proceedings of symposia related to cancer research.
5) Announcements of future meetings of interest to readers: courses in cancer-related biomedical science, or the availability of fellowships; and listings of relevant books and other publications. Announcements should be submitted at least six (6) months prior to date of issue.

All papers and other submissions will be sent to one or more editors for peer review. It is our policy to return those papers which do not meet the requirements set forth in these instructions.

Manuscripts should be sent to *NUTRITION AND CANCER: AN INTERNATIONAL JOURNAL*, c/o Kathy Dolan, Lawrence Erlbaum Associates, 10 Industrial Ave., Mahwah, NJ 07430–2262. The address to which page proofs and correspondence should be sent must be specified. Include phone, fax, and e-mail information. Include verification that the article has not been submitted concurrently to any other journals.

Manuscripts will not be returned. Manuscripts should be typed on 8½ × 11" paper with double spacing throughout, allowing for ample margins. Submit one (1) original and five (5) copies. The manuscript should be organized in the following manner: title page, abstract, text, acknowledgments and notes, references, appendixes, tables, figure captions, copies of all figures, and two sets of original figures. Also submit two (2) sets of complete illustrations. Consecutive numbering of all pages is requested, with the title page as page one. Title page should provide the name of author and co-authors. The first author's last name plus the page number belong in the upper right corner of each page.

Authors are urged to include their full names, complete with first and middle initials to avoid confusion which often arises when authors are identified by surname and initials only. Authors' academic degrees should not be included. The full names of institutions and subsidiary laboratories should be given, along with a useful address (including postal number). If several authors (maximum 10 authors) and institutions are listed on a paper, it should be clearly indicated with which institution each author is affiliated.

For text style, authors should follow *Scientific Style and Format: The CBE Manual for Authors, Editors, and Publishers* (6th edition, 1994) in matters of spelling, capitalization, punctuation, hyphenation, and general style; *Current Procedural Terminology* and *International Classification of Diseases* for terms relating to diseases, operations, and procedures; *Chemical Abstracts* for chemical terms; and *Serial Sources for the BIOSIS Previews Database* for journal abbreviations in references. Metric equivalents are preferred.

The list of references should be typed double-spaced and numbered consecutively as they appear. List only five (5) authors before et al. Authors are responsible for accuracy and must check every reference in manuscript and proofread again in page proof.

Journal references should be given in the following order: author, article, title, journal abbreviation, volume number in Arabic numerals, inclusive pages and year. If the paper has been seen only in abstract, this should be indicated at the end of the original reference by the addition of the abbreviation (abstr), followed by the abstracting source (including volume, page, and year). The order for book references is as follows: author, title, edition number (if other than the first), city, publisher, year, and volume (if more than one). If the reference is a chapter in a book, the order changes as follows: author of the chapter, title of the chapter, book title, edition, editor(s), city, publisher, year, and inclusive pages of the chapter.

The hospital or academic institution and city where the work was done; the source of financial support; an acknowledgment (if desired) of those who aided in research and preparation of the manuscript; and a mailing address for reprints (if available) should appear at the end of the text (before references). All trade names of drugs should be referenced with the generic name and the name, city, and state of the manufacturer.

Line art accompanying papers should be supplied as clear, black-and-white printouts on 8½ × 11" bond paper. Photographs should be mounted on plates of white cardboard; the overall dimensions of photographs on a given plate should not exceed 18.4 × 22.4 cm (7¼ × 9"). All photographs should be correctly exposed, sharply focused, and submitted on glossy white paper. Photographed electrophoretic patterns should be as small as possible to avoid the need for returning them to the author for reduction. Do not submit bar graphs with more than one shade of gray; gray tint must be easily distinguishable from black tint. Patterns (hatched, cross hatched, dots) or solid black and white are preferable to gray tints.

Because halftone illustrations (photomicrographs and photographs) are expensive to reproduce, only those photographs that are absolutely essential to the clarity of the presentation can be accepted.

Color photographs can be published only if after careful evaluation they are accepted by the editors and the complete expense of reproducing and tipping in the color pages is received in full in advance from the author. Current estimates for color reproduction can be obtained by corresponding with the publisher.

Figure legends should be typed on a separate sheet, double-spaced in consecutive order. The author must obtain permission from the previous author and copyright holder to reproduce illustrations published previously and acknowledgment should be listed in the legend.

Several stylistic items are commonly overlooked by authors, thus entailing wasted time and expense at the processing and publication stages. Authors may find it helpful to refer to the following checklist before transmitting manuscripts to our office. 1) Manuscript submitted with five copies (include tables and figures) and two sets of original illustrations. 2) Grant information, if appropriate. 3) Exact affiliation of each author given. 4) Abstract included (200 words maximum). 5) All nonstandard abbreviations defined. 6) Exact location (city and state, or country) supplied for sources of special chemicals or preparations. 7) All references listed in order of appearance and typed double-spaced.

Reprint request forms, mailed to authors with page proofs, *must* be accompanied by proofs. Reprints are offered only at this time.